Practical Data Analysis Cookbook

Over 60 practical recipes on data exploration and analysis

Tomasz Drabas

[PACKT] open source *
PUBLISHING   community experience distilled

BIRMINGHAM - MUMBAI

Practical Data Analysis Cookbook

First published: April 2011

Production reference: 1250416

Published by Packt Publishing Ltd.
Livery Place
35 Livery Street
Birmingham B3 2PB, UK.

ISBN 978-1-78355-166-8

www.packtpub.com

Credits

Author
Tomasz Drabas

Reviewers
Brett Bloomquist

Khaled Tannir

Commissioning Editor
Dipika Gaonkar

Acquisition Editor
Prachi Bisht

Content Development Editor
Pooja Mhapsekar

Technical Editor
Bharat Patil

Copy Editor
Tasneem Fatehi

Project Coordinator
Francina Pinto

Proofreader
Safis Editing

Indexer
Mariammal Chettiyar

Graphics
Name

Production Coordinator
Nilesh R. Mohite

Cover Work
Nilesh R. Mohite

About the Author

Tomasz Drabas is a data scientist working for Microsoft and currently residing in the Seattle area. He has over 12 years of international experience in data analytics and data science in numerous fields, such as advanced technology, airlines, telecommunications, finance, and consulting.

Tomasz started his career in 2003 with LOT Polish Airlines in Warsaw, Poland, while finishing his master's degree in strategy management. In 2007, he moved to Sydney to pursue a doctoral degree in operations research at the University of New South Wales, School of Aviation; his research crossed boundaries between discrete choice modeling and airline operations research. During his time in Sydney, he worked as a data analyst for Beyond Analysis Australia and as a senior data analyst/data scientist for Vodafone Hutchison Australia, among others. He has also published scientific papers, attended international conferences, and served as a reviewer for scientific journals.

In 2015, he relocated to Seattle to begin his work for Microsoft. There he works on numerous projects involving solving problems in high-dimensional feature space.

Acknowledgments

First and foremost, I would like to thank my wife, Rachel, and daughter, Skye, for encouraging me to undertake this challenge and tolerating long days of developing code and late nights of writing up. You are the best and I love you beyond bounds! Also, thanks to my family for putting up with me (in general).

Tomasz Bednarz has not only been a great friend but also a great mentor when I was learning programming—thank you! I also want to thank my current and former managers, Mike Stephenson and Rory Carter, as well as numerous colleagues and friends who also encouraged me to finish this book.

Special thanks go to my two former supervisors, Dr Richard Cheng-Lung Wu and Dr Tomasz Jablonski. The master's project with Tomasz sparked my interest in neural networks—lessons that I will never forget. Without Richard's help, I would not have been able to finish my PhD and will always be grateful for his help, guidance, and friendship.

About the Reviewers

Brett Bloomquist holds a BS in mathematics and an MS in computer science, specializing in computer-aided geometric design. He has 26 years of work experience in the software industry with a focus on geometric modeling algorithms and computer graphics. More recently, Brett has been applying his mathematics and visualization background as a principal data scientist.

Khaled Tannir is a visionary solution architect with more than 20 years of technical experience focusing on big data technologies, data science machine learning, and data mining since 2010.

He is widely recognized as an expert in these fields and has a bachelor's degree in electronics and a master's degree in system information architectures. He is working on completing his PhD.

Khaled has more than 15 certifications (R programming, big data, and many more) and is a Microsoft Certified Solution Developer (MCSD) and an avid technologist.

He has worked for many companies in France (and recently in Canada), leading the development and implementation of software solutions and giving technical presentations.

He is the author of the books *RavenDB 2.x Beginner's Guide* and *Optimizing Hadoop MapReduce*, both by Packt Publishing (which were translated in Simplified Chinese) and a technical reviewer on the books, *Pentaho Analytics for MongoDB*, *MongoDB High Availability*, and *Learning Predictive Analytics with R*, by Packt Publishing.

He enjoys taking landscape and night photos, traveling, playing video games, creating funny electronics gadgets using Arduino, Raspberry Pi, and .Net Gadgeteer, and of course spending time with his wife and family.

You can connect with him on LinkedIn or reach him at `contact@khaledtannir.net`.

www.PacktPub.com

Support files, eBooks, discount offers, and more

For support files and downloads related to your book, please visit www.PacktPub.com.

Did you know that Packt offers eBook versions of every book published, with PDF and ePub files available? You can upgrade to the eBook version at www.PacktPub.com and as a print book customer, you are entitled to a discount on the eBook copy. Get in touch with us at service@packtpub.com for more details.

At www.PacktPub.com, you can also read a collection of free technical articles, sign up for a range of free newsletters and receive exclusive discounts and offers on Packt books and eBooks.

https://www2.packtpub.com/books/subscription/packtlib

Do you need instant solutions to your IT questions? PacktLib is Packt's online digital book library. Here, you can search, access, and read Packt's entire library of books.

Why Subscribe?

- ► Fully searchable across every book published by Packt
- ► Copy and paste, print, and bookmark content
- ► On demand and accessible via a web browser

Free Access for Packt account holders

If you have an account with Packt at www.PacktPub.com, you can use this to access PacktLib today and view 9 entirely free books. Simply use your login credentials for immediate access.

Table of Contents

Preface

Data analytics and data science have garnered a lot of attention from businesses around the world. The amount of data generated these days is mind-boggling, and it keeps growing everyday; with the proliferation of mobiles, access to Facebook, YouTube, Netflix, or other 4K video content providers, and increasing reliance on cloud computing, we can only expect this to increase.

The task of a data scientist is to clean, transform, and analyze the data in order to provide the business with insights about its customers and/or competitors, monitor the health of the services provided by the company, or automatically present recommendations to drive more opportunities for cross-selling (among many others).

In this book, you will learn how to read, write, clean, and transform the data—the tasks that are the most time-consuming but also the most critical. We will then present you with a broad array of tools and techniques that any data scientist should master, ranging from classification, clustering, or regression, through graph theory and time-series analysis, to discrete choice modeling and simulations. In each chapter, we will present an array of detailed examples written in Python that will help you tackle virtually any problem that you might encounter in your career as a data scientist.

What this book covers

Chapter 1, Preparing the Data, covers the process of reading and writing from and to various data formats and databases, as well as cleaning the data using OpenRefine and Python.

Chapter 2, Exploring the Data, describes various techniques that aid in understanding the data. We will see how to calculate distributions of variables and correlations between them and produce some informative charts.

Chapter 3, Classification Techniques, introduces several classification techniques, from simple Naïve Bayes classifiers to more sophisticated Neural Networks and Random Tree Forests.

Chapter 4, *Clustering Techniques*, explains numerous clustering models; we start with the most common k-means method and finish with more advanced BIRCH and DBSCAN models.

Chapter 5, *Reducing Dimensions*, presents multiple dimensionality reduction techniques, starting with the most renowned PCA, through its kernel and randomized versions, to LDA.

Chapter 6, *Regression Methods*, covers many regression models, both linear and nonlinear. We also bring back random forests and SVMs (among others) as these can be used to solve either classification or regression problems.

Chapter 7, *Time Series Techniques*, explores the methods of handling and understanding time series data as well as building ARMA and ARIMA models.

Chapter 8, *Graphs*, introduces NetworkX and Gephi to handle, understand, visualize, and analyze data in the form of graphs.

Chapter 9, *Natural Language Processing*, describes various techniques related to the analytics of free-flow text: part-of-speech tagging, topic extraction, and classification of data in textual form.

Chapter 10, *Discrete Choice Models*, explains the choice modeling theory and some of the most popular models: the Multinomial, Nested, and Mixed Logit models.

Chapter 11, *Simulations*, covers the concepts of agent-based simulations; we simulate the functioning of a gas station, out-of-power occurrences for electric vehicles, and sheep-wolf predation scenarios.

What you need for this book

For this book, you need a personal computer (it can be a Windows machine, Mac, or Linux) with an installed and configured Python 3.5 environment; we use the Anaconda distribution of Python that can be downloaded at `https://www.continuum.io/downloads`.

Throughout this book, we use various Python modules: pandas, NumPy/SciPy, SciKit-Learn, MLPY, StatsModels, PyBrain, NLTK, BeautifulSoup, Optunity, Matplotlib, Seaborn, Bokeh, PyLab, OpenPyXl, PyMongo, SQLAlchemy, NetworkX, and SimPy. Most of the modules used come preinstalled with Anaconda, but some of them need to be installed via either the `conda` installer or by downloading the module and using the `python setup.py install` command. It is fine if some of those modules are not currently installed on your machine; we will guide you through the installation process.

We also use several non-Python tools: OpenRefine to aid in data cleansing and analysis, D3.js to visualize data, Postgres and MongoDB databases to store data, Gephi to visualize graphs, and PythonBiogeme to estimate discrete choice models. We will provide detailed installation instructions where needed.

Who this book is for

This book is for everyone who wants to get into the data science field and needs to build up their skills on a set of examples that aim to tackle the problems faced in the corporate world. More advanced practitioners might also find some of the examples refreshing and the more advanced topics covered interesting.

Sections

In this book, you will find several headings that appear frequently (Getting ready, How to do it, How it works, There's more, and See also).

To give clear instructions on how to complete a recipe, we use these sections as follows:

Getting ready

This section tells you what to expect in the recipe, and describes how to set up any software or any preliminary settings required for the recipe.

How to do it...

This section contains the steps required to follow the recipe.

How it works...

This section usually consists of a detailed explanation of what happened in the previous section.

There's more...

This section consists of additional information about the recipe in order to make the reader more knowledgeable about the recipe.

See also

This section provides helpful links to other useful information for the recipe.

Conventions

In this book, you will find a number of text styles that distinguish between different kinds of information. Here are some examples of these styles and an explanation of their meaning.

Code words in text, database table names, folder names, filenames, file extensions, pathnames, dummy URLs, user input, and Twitter handles are shown as follows: "We can include other contexts through the use of the `include` directive."

A block of code is set as follows:

```
for p in all_disputed_transactions:
    try:
        transactions[p[0]].append(p[2]['amount'])
    except:
        transactions[p[0]] = [p[2]['amount']]
```

Any command-line input or output is written as follows:

```
cd networkx
python setup.py install
```

New terms and **important words** are shown in bold. Words that you see on the screen, for example, in menus or dialog boxes, appear in the text like this: "We start with using **Range** on the age filter."

> Warnings or important notes appear in a box like this.

> Tips and tricks appear like this.

Reader feedback

Feedback from our readers is always welcome. Let us know what you think about this book—what you liked or disliked. Reader feedback is important for us as it helps us develop titles that you will really get the most out of.

To send us general feedback, simply e-mail feedback@packtpub.com, and mention the book's title in the subject of your message.

If there is a topic that you have expertise in and you are interested in either writing or contributing to a book, see our author guide at www.packtpub.com/authors.

Customer support

Now that you are the proud owner of a Packt book, we have a number of things to help you to get the most from your purchase.

Downloading the example code

You can download the example code files from your account at `http://www.packtpub.com` for all the Packt Publishing books you have purchased. If you purchased this book elsewhere, you can visit `http://www.packtpub.com/support` and register to have the files e-mailed directly to you.

You can download the code files by following these steps:

1. Log in or register to our website using your e-mail address and password.
2. Hover the mouse pointer on the **SUPPORT** tab at the top.
3. Click on **Code Downloads & Errata**.
4. Enter the name of the book in the **Search** box.
5. Select the book for which you're looking to download the code files.
6. Choose from the drop-down menu where you purchased this book from.
7. Click on **Code Download**.

You can also download the code files by clicking on the **Code Files** button on the book's webpage at the Packt Publishing website. This page can be accessed by entering the book's name in the **Search** box. Please note that you need to be logged in to your Packt account.

Once the file is downloaded, please make sure that you unzip or extract the folder using the latest version of:

- WinRAR / 7-Zip for Windows
- Zipeg / iZip / UnRarX for Mac
- 7-Zip / PeaZip for Linux

The code bundle for this book is also available on GitHub at `https://github.com/drabastomek/practicalDataAnalysisCookbook/tree/master/Data`.

Downloading the color images of this book

We also provide you with a PDF file that has color images of the screenshots/diagrams used in this book. The color images will help you better understand the changes in the output. You can download this file from `https://www.packtpub.com/sites/default/files/downloads/practicaldataanalysiscookbook_ColorImages.pdf`.

Errata

Although we have taken every care to ensure the accuracy of our content, mistakes do happen. If you find a mistake in one of our books—maybe a mistake in the text or the code—we would be grateful if you could report this to us. By doing so, you can save other readers from frustration and help us improve subsequent versions of this book. If you find any errata, please report them by visiting `http://www.packtpub.com/submit-errata`, selecting your book, clicking on the **Errata Submission Form** link, and entering the details of your errata. Once your errata are verified, your submission will be accepted and the errata will be uploaded to our website or added to any list of existing errata under the Errata section of that title.

To view the previously submitted errata, go to `https://www.packtpub.com/books/content/support` and enter the name of the book in the search field. The required information will appear under the **Errata** section.

Piracy

Piracy of copyrighted material on the Internet is an ongoing problem across all media. At Packt, we take the protection of our copyright and licenses very seriously. If you come across any illegal copies of our works in any form on the Internet, please provide us with the location address or website name immediately so that we can pursue a remedy.

Please contact us at `copyright@packtpub.com` with a link to the suspected pirated material.

We appreciate your help in protecting our authors and our ability to bring you valuable content.

Questions

If you have a problem with any aspect of this book, you can contact us at `questions@packtpub.com`, and we will do our best to address the problem.

1
Preparing the Data

In this chapter, we will cover the basic tasks of reading, storing, and cleaning data using Python and OpenRefine. You will learn the following recipes:

- Reading and writing CSV/TSV files with Python
- Reading and writing JSON files with Python
- Reading and writing Excel files with Python
- Reading and writing XML files with Python
- Retrieving HTML pages with pandas
- Storing and retrieving from a relational database
- Storing and retrieving from MongoDB
- Opening and transforming data with OpenRefine
- Exploring the data with OpenRefine
- Removing duplicates
- Using regular expressions and GREL to clean up the data
- Imputing missing observations
- Normalizing and standardizing features
- Binning the observations
- Encoding categorical variables

Introduction

For the following set of recipes, we will use Python to read data in various formats and store it in RDBMS and NoSQL databases.

All the source codes and datasets that we will use in this book are available in the GitHub repository for this book. To clone the repository, open your terminal of choice (on Windows, you can use command line, Cygwin, or Git Bash and in the Linux/Mac environment, you can go to Terminal) and issue the following command (in one line):

```
git clone https://github.com/drabastomek/practicalDataAnalysisCookbook.
git
```

 Note that you need Git installed on your machine. Refer to `https://git-scm.com/book/en/v2/Getting-Started-Installing-Git` for installation instructions.

In the following four sections, we will use a dataset that consists of 985 real estate transactions. The real estate sales took place in the Sacramento area over a period of five consecutive days. We downloaded the data from `https://support.spatialkey.com/spatialkey-sample-csv-data/`—in specificity, `http://samplecsvs.s3.amazonaws.com/Sacramentorealestatetransactions.csv`. The data was then transformed into various formats that are stored in the `Data/Chapter01` folder in the GitHub repository.

In addition, you will learn how to retrieve information from HTML files. For this purpose, we will use the Wikipedia list of airports starting with the letter A, `https://en.wikipedia.org/wiki/List_of_airports_by_IATA_code:_A`.

To clean our dataset, we will use OpenRefine; it is a powerful tool to read, clean, and transform data.

Reading and writing CSV/TSV files with Python

CSV and TSV formats are essentially text files formatted in a specific way: the former one separates data using a comma and the latter uses tab `\t` characters. Thanks to this, they are really portable and facilitate the ease of sharing data between various platforms.

Getting ready

To execute this recipe, you will need the `pandas` module installed. These modules are all available in the Anaconda distribution of Python and no further work is required if you already use this distribution. Otherwise, you will need to install `pandas` and make sure that it loads properly.

You can download Anaconda from `http://docs.continuum.io/anaconda/install`. If you already have Python installed but do not have `pandas`, you can download the package from `https://github.com/pydata/pandas/releases/tag/v0.17.1` and follow the instructions to install it appropriately for your operating system (`http://pandas.pydata.org/pandas-docs/stable/install.html`).

No other prerequisites are required.

How to do it...

The `pandas` module is a library that provides high-performing, high-level data structures (such as DataFrame) and some basic analytics tools for Python.

 The DataFrame is an Excel table-like data structure where each column represents a feature of your dataset (for example, the height and weight of people) and each row holds the data (for example, 1,000 random people's heights and weights). See `http://pandas.pydata.org/pandas-docs/stable/dsintro.html#dataframe`.

The module provides methods that make it very easy to read data stored in a variety of formats. Here's a snippet of a code that reads the data from CSV and TSV formats, stores it in a `pandas` DataFrame structure, and then writes it back to the disk (the `read_csv.py` file):

```python
import pandas as pd

# names of files to read from
r_filenameCSV = '../../Data/Chapter01/realEstate_trans.csv'
r_filenameTSV = '../../Data/Chapter01/realEstate_trans.tsv'

# names of files to write to
w_filenameCSV = '../../Data/Chapter01/realEstate_trans.csv'
w_filenameTSV = '../../Data/Chapter01/realEstate_trans.tsv'

# read the data
csv_read = pd.read_csv(r_filenameCSV)
tsv_read = pd.read_csv(r_filenameTSV, sep='\t')

# print the first 10 records
print(csv_read.head(10))
print(tsv_read.head(10))
```

```
# write to files
with open(w_filenameCSV,'w') as write_csv:
    write_csv.write(tsv_read.to_csv(sep=',', index=False))

with open(w_filenameTSV,'w') as write_tsv:
    write_tsv.write(csv_read.to_csv(sep='\t', index=False))
```

Now, open the command-line console (on Windows, you can use either command line or Cygwin and in the Linux/Mac environment, you go to Terminal) and execute the following command:

python read_csv.py

You shall see an output similar to the following (abbreviated):

```
Baths beds       city      latitude   longitude    price  \
0     1     2   SACRAMENTO  38.631913  -121.434879  59222
1     1     3   SACRAMENTO  38.478902  -121.431028  68212
2     1     2   SACRAMENTO  38.618305  -121.443839  68880
...
```

How it works...

First, we load `pandas` to get access to the DataFrame and all its methods that we will use to read and write the data. Note that we alias the `pandas` module using `as` and specifying the name, `pd`; we do this so that later in the code we do not need to write the full name of the package when we want to access DataFrame or the `read_csv(...)` method. We store the filenames (for the reading and writing) in `r_filenameCSV(TSV)` and `w_filenameCSV(TSV)` respectively.

To read the data, we use `pandas'` `read_csv(...)` method. The method is very universal and accepts a variety of input parameters. However, at a minimum, the only required parameter is either the filename of the file or a buffer that is, an opened file object. In order to read the `realEstate_trans.tsv` file, you might want to specify the `sep='\t'` parameter; by default, `read_csv(...)` will try to infer the separator but I do not like to leave it to chance and always specify the separator explicitly.

As the two files hold exactly the same data, you can check whether the files were read properly by printing out some records. This can be accomplished using the `.head(<no_of_rows>)` method invoked on the DataFrame object, where `<no_of_rows>` specifies how many rows to print out.

Storing the data in `pandas'` DataFrame object means that it really does not matter what format the data was initially in; once read, it can then be saved in any format supported by `pandas`. In the preceding example, we write the contents read from a CSV file to a file in a TSV format.

The `with open(...) as ...:` structure should always be used to open files for either reading or writing. The advantage of opening files in this way is that it closes the file properly once you are done with reading from or writing to even if, for some reason, an exception occurs during the process.

An exception is a situation that the programmer did not expect to see when he or she wrote the program.

Consider, for example, that you have a file where each line contains only one number: you open the file and start reading from it. As each line of the file is treated as text when read, you need to transform the read text into an integer—a data structure that a computer understands (and treats) as a number, not a text.

All is fine if your data really contains only numbers. However, as you will learn later in this chapter, all data that we gather is dirty in some way, so if, for instance, any of the rows contains a letter instead of a number, the transformation will fail and Python will raise an exception.

The open(<filename>, 'w') command opens the file specified by <filename> to write (the w parameter). Also, you can open files in read mode by specifying 'r' instead. If you open a file in the 'r+' mode, Python will allow a bi-directional flow of data (read and write) so you will be able to append contents at the end of the file if needed. You can also specify rb or wb for binary type of data (not text).

The .to_csv(...) method converts the content of a DataFrame to a format ready to store in a text file. You need to specify the separator, for example, sep=', ', and whether the DataFrame index is to be stored in the file as well; by default, the index is also stored. As we do not want that, you should specify index=False.

 The DataFrame's index is essentially an easy way to identify, align, and access your data in the DataFrame. The index can be a consecutive list of numbers (just like row numbers in Excel) or dates; you can even specify two or more index columns. The index column is not part of your data (even though it is printed to screen when you print the DataFrame object). You can read more about indexing at http://pandas.pydata.org/pandas-docs/ stable/indexing.html.

There's more...

Described here is the easiest and quickest way of reading data from and writing data to CSV and TSV files. If you prefer to hold your data in a data structure other than pandas' DataFrame, you can use the csv module. You then read the data as follows (the read_csv_alternative.py file):

```python
import csv

# names of files to read from
r_filenameCSV = '../../Data/Chapter01/realEstate_trans.csv'
r_filenameTSV = '../../Data/Chapter01/realEstate_trans.tsv'

# data structures to hold the data
csv_labels = []
tsv_labels = []
csv_data = []
tsv_data = []

# read the data
with open(r_filenameCSV, 'r') as csv_in:
    csv_reader = csv.reader(csv_in)

    # read the first line that holds column labels
```

```
        csv_labels = csv_reader.__next__()

        # iterate through all the records
        for record in csv_reader:
            csv_data.append(record)

    with open(r_filenameTSV, 'r') as tsv_in:
        tsv_reader = csv.reader(tsv_in, delimiter='\t')

        tsv_labels = tsv_reader.__next__()

        for record in tsv_reader:
            tsv_data.append(record)

    # print the labels
    print(csv_labels, '\n')
    print(tsv_labels, '\n')

    # print the first 10 records
    print(csv_data[0:10],'\n')
    print(tsv_data[0:10],'\n')
```

We store the labels and data in separate lists, `csv(tsv)_labels` and `csv(tsv)_data` respectively. The `.reader(...)` method reads the data from the specified file line by line. To create a `.reader(...)` object, you need to pass an open CSV or TSV file object. In addition, if you want to read a TSV file, you need to specify the delimiter as well, just like DataFrame.

 The `csv` module also provides the `csv.writer` object that allows saving data in a CSV/TSV format. See the documentation of the `csv` module at `https://docs.python.org/3/library/csv.html`.

See also

Check the `pandas` documentation for `read_csv(...)` and `write_csv(...)` to learn more about the plethora of parameters these methods accept. The documentation can be found at `http://pandas.pydata.org/pandas-docs/stable/io.html#io-read-csv-table`.

Reading and writing JSON files with Python

JSON stands for **JavaScript Object Notation**. It is a hierarchical dictionary-like structure that stores key-value pairs separated by a comma; the key-value pairs are separated by a colon ': '. JSON is platform-independent (like XML, which we will cover in the *Reading and writing XML files with Python* recipe) making sharing data between platforms very easy. You can read more about JSON at http://www.w3schools.com/json/.

Getting ready

To execute this recipe, you will need Python with the `pandas` module installed. No other prerequisites are required.

How to do it...

The code to read a JSON file is as follows. Note that we assume the `pandas` module is already imported and aliased as pd (the `read_json.py` file):

```
# name of the JSON file to read from
r_filenameJSON = '../../Data/Chapter01/realEstate_trans.json'

# read the data
json_read = pd.read_json(r_filenameJSON)

# print the first 10 records
print(json_read.head(10))
```

How it works...

This code works in a similar way to the one introduced in the previous section. First, you need to specify the name of the JSON file—we store it in the `r_filenameJSON` string. Next, use the `read_json(...)` method of pandas, passing `r_filenameJSON` as the only parameter.

The read data is stored in the `json_read DataFrame` object. We then print the bottom 10 observations using the `.tail(...)` method. To write a JSON file, you can use the `.to_json()` method on DataFrame and write the returned data to a file in a similar manner as discussed in the *Reading and writing CSV/TSV files with Python* recipe.

There's more...

You can read and write JSON files using the `json` module as well. To read data from a JSON file, you can refer to the following code (the `read_json_alternative.py` file):

```
# read the data
with open('../../Data/Chapter01/realEstate_trans.json', 'r') \
    as json_file:
        json_read = json.loads(json_file.read())
```

This code reads the data from the `realEstate_trans.json` file and stores it in a `json_read` list. It uses the `.read()` method on a file that reads the whole content of the specified file into memory. To store the data in a JSON file, you can use the following code:

```
# write back to the file
with open('../../Data/Chapter01/realEstate_trans.json', 'w') \
    as json_file:
        json_file.write(json.dumps(json_read))
```

See also

Check the `pandas` documentation for `read_json` at `http://pandas.pydata.org/pandas-docs/stable/io.html#io-json-reader`.

Reading and writing Excel files with Python

Microsoft Excel files are arguably the most widely used format to exchange data in a tabular form. In the newest incarnation of the `XLSX` format, Excel can store over one million rows and over 16 thousand columns in a single worksheet.

Getting ready

To execute this recipe, you will need the `pandas` module installed. No other prerequisites are required.

How to do it...

The following is the code to read the Excel file. Note that we assume the `pandas` module is already imported and aliased as `pd` (the `read_xlsx.py` file):

```
# name of files to read from and write to
r_filenameXLSX = '../../Data/Chapter01/realEstate_trans.xlsx'
w_filenameXLSX = '../../Data/Chapter01/realEstate_trans.xlsx'
```

```
# open the Excel file
xlsx_file = pd.ExcelFile(r_filenameXLSX)

# read the contents
xlsx_read = {
    sheetName: xlsx_file.parse(sheetName)
        for sheetName in xlsx_file.sheet_names
}

# print the first 10 prices for Sacramento
print(xlsx_read['Sacramento'].head(10)['price'])

# write to Excel
xlsx_read['Sacramento'] \
    .to_excel(w_filenameXLSX, 'Sacramento', index=False)
```

How it works...

We follow a similar manner to the previous examples. We first open the XLSX file and assign it to the `xlsx_file` object using `pandas'` `ExcelFile(...)` method. We employ the `.parse(...)` method to do the work for us and read the contents of the specified worksheet; we store it in the `xlsx_read` dictionary. Note that you get access to all the worksheets in the Excel file through the `.sheet_names` property of the `ExcelFile` object.

To create the `xlsx_read` dictionary, we use Pythonic dictionary comprehension: instead of looping through the sheets explicitly and then adding the elements to the dictionary, we use the dictionary comprehension to make the code more readable and compact.

The comprehensions make it easy to understand the code as they mimic mathematical notations. Consider, for example, the following list of powers of 2: ($A = (2^0, 2^1, 2^2, ..., 2^8)$ $= (2^x: 0 <= x < 9)$, x is an integer). It can then easily be translated into Python using a list comprehension: `A = [2**x for x in range(0, 9)]`. This would create the following list: `A = [1, 2, 4, 8, 16, 32, 64, 128, 256]`.

Also, in Python, the comprehensions are also a tiny bit faster than explicit loops (`http://stackoverflow.com/questions/22108488/are-list-comprehensions-and-functional-functions-faster-than-for-loops`).

The `range(<from>,<to>)` command generates a sequence of integers starting at `<from>` and extending to `<to>` less one. For example, `range(0,3)` will generate a sequence 0, 1, 2.

Storing the data in an Excel file is also very easy. All that is required is to invoke the `.to_excel(...)` method, where the first parameter is the name of the file you want to save the data to and the second one specifies the name of the worksheet. In our example, we also specified the additional `index=False` parameter that instructs the method not to save the index; by default, the `.to_excel(...)` method saves the index in column A.

There's more...

Alternatively to reading Excel files using `pandas' read_excel(...)`, there are multiple Python modules you can use that provide Excel data reading capabilities. `pandas` uses the `xlrd` (https://secure.simplistix.co.uk/svn/xlrd/trunk/xlrd/doc/xlrd.html?p=4966) module to read the data and then converts it to a DataFrame. For XLSX files, you can also use the `openpyxl` module (the `read_xlsx_alternative.py` file):

```
import openpyxl as oxl

# name of files to read from
r_filenameXLSX = '../../Data/Chapter01/realEstate_trans.xlsx'

# open the Excel file
xlsx_wb = oxl.load_workbook(filename=r_filenameXLSX)

# names of all the sheets in the workbook
sheets = xlsx_wb.get_sheet_names()

# extract the 'Sacramento' worksheet
xlsx_ws = xlsx_wb[sheets[0]]
```

We first read the contents of the Excel file and store it in `xlsx_wb` (workbook). From the workbook, we extract the names of all the worksheets and put it in the `sheets` variable. As we have only one worksheet in our workbook, the `sheets` variable equals to `'Sacramento'`. We use it to create an `xlsx_ws` object that allows iterating through all the rows:

```
labels = [cell.value for cell in xlsx_ws.rows[0]]

data = [] # list to hold the data

for row in xlsx_ws.rows[1:]:
    data.append([cell.value for cell in row])
```

The first row contains the labels for all the columns so it is a good idea to store this separately—we put it in the `labels` variable. We then iterate through all the rows in the worksheet, using the `.rows` iterator, and append the values of all the cells to the `data` list:

```
print(
    [item[labels.index('price')] for item in data[0:10]]
)
```

The last part of the code prints out the prices of properties for the top 10 rows. We use list comprehension to create a list of the prices. You can find the first occurrence of a certain item in a list by calling `.index(...)` on a list object, as we did in this example.

See also

Check the `pandas` documentation for `read_excel` at `http://pandas.pydata.org/pandas-docs/stable/io.html#io-excel`. Also, you can visit `http://www.python-excel.org` for a list of modules that allow you to work with data stored in different Excel formats, both older `.xls` and newer `.xlsx` files.

Reading and writing XML files with Python

XML stands for eXtensible Markup Language. Although not as popular to store data as the formats described previously, certain web APIs return XML-encoded information on request.

An XML-encoded document has a tree-like structure. To read the contents, we start at the root of the tree (normally, the name of the element that follows the XML declaration `<?xml version="1.0" encoding="UTF-8"?>`; every XML-encoded document needs to begin with such declaration). In our case, the root of our XML-encoded document is `<records>`. A single `<record>...</record>` contains a list of `<var var_name=...>...</var>`.

> Warning: The `xml` module is not secure. Caution is required when dealing with XML-encoded messages from untrusted sources. An attacker might access local files, carry out DoS attacks, and more. Refer to the documentation for the `xml` module at `https://docs.python.org/3/library/xml.html`.

Getting ready

In order to execute the following recipe, you need the `pandas` and `xml` modules available. No other prerequisites are required.

How to do it...

Reading the data from an XML file directly to a `pandas` DataFrame requires some supplementary code; this is because each XML file has a different structure and requires a made-to-fit parsing. We will define the innards of the methods defined in the following section of this recipe. The source code for this section can be found in the `read_xml.py` file:

```
import pandas as pd
import xml.etree.ElementTree as ET
```

```
def read_xml(xml_tree):
    '''
        Read an XML encoded data and return pd.DataFrame
    '''

def iter_records(records):
    '''
        Generator to iterate through all the records
    '''

def write_xml(xmlFileName, data):
    '''
        Save the data in an XML format
    '''

def xml_encode(row):
    '''
        Encode the row as an XML with a specific hierarchy
    '''

# names of files to read from and write to
r_filenameXML = '../../Data/Chapter01/realEstate_trans.xml'
w_filenameXML = '../../Data/Chapter01/realEstate_trans.xml'

# read the data
xml_read = read_xml(r_filenameXML)

# print the first 10 records
print(xml_read.head(10))

# write back to the file in an XML format
write_xml(w_filenameXML, xml_read)
```

How it works...

Let's analyze the preceding code step by step. First, we import all the modules that we need. The xml.etree.ElementTree module is a lightweight XML parser of the XML tree and we will use it to parse the XML structure of our file. As before, we define names of the files to read and write in separate variables (r_filenameXML, w_filenameXML).

To read the data from the XML-encoded file, we use the read_xml(...) method:

```
def read_xml(xmlFileName):
    with open(xmlFileName, 'r') as xml_file:
        # read the data and store it as a tree
```

```
tree = ET.parse(xml_file)

# get the root of the tree
root = tree.getroot()

# return the DataFrame
return pd.DataFrame(list(iter_records(root)))
```

The method takes the name of the file as its only parameter. First, the file is opened. Using the `.parse(...)` method, we create a tree-like structure from our XML-encoded file and store it in the `tree` object. We then extract the root using the `.getroot()` method on the `tree` object: this is the starting point to process the data further. The return statement calls the `iter_records` method passing the reference to the root of the tree and then converts the returned information to a DataFrame:

```
def iter_records(records):
    for record in records:
        # temporary dictionary to hold values
        temp_dict = {}

        # iterate through all the fields
        for var in record:
            temp_dict[
                var.attrib['var_name']
            ] = var.text

        # generate the value
        yield temp_dict
```

The `iter_records` method is a *generator*: a method that, as the name suggests, generates the values. Unlike regular methods that have to return all the values when the function finishes (a `return` statement), generators hand over the data back to the calling method one at a time (hence the `yield` keyword) until done.

> For a more in-depth discussion on generators, I suggest reading `https://www.jeffknupp.com/blog/2013/04/07/improve-your-python-yield-and-generators-explained/`.

Our `iter_records` method, for each record read, emits a `temp_dict` dictionary object back to the `read_xml` method. Each element of the dictionary has a key equal to the `var_name` attribute of the `<var>` XML element. (Our `<var>` has the following format: `<var var_name=...>`.)

The `<var>` tag could have more attributes with other names—these would be stored in the `.attrib` dictionary (a property of the XML tree node) and would be accessible by their names—see the highlighted line in the previous source code.

The value of `<var>` (contained within `<var>...</var>`) is accessible through the `.text` property of the XML node, while the `.tag` property stores its name (in our case, `var`).

The `return` statement of the `read_xml` method creates a list from all the dictionaries passed, which is then turned into a DataFrame.

To write the data in an XML format, we use the `write_xml(...)` method:

```
def write_xml(xmlFileName, data):
    with open(xmlFileName, 'w') as xmlFile:

        # write the headers
        xmlFile.write(
            '<?xml version="1.0" encoding="UTF-8"?>\n'
        )
        xmlFile.write('<records>\n')

        # write the data
        xmlFile.write(
            '\n'.join(data.apply(xml_encode, axis=1))
        )

        # write the footer
        xmlFile.write('\n</records>')
```

The method opens the file specified by the `xmlFileName` parameter. Every XML file needs to start with the XML declaration (see the introduction to this recipe) in the first line. Then, we write out the root of our XML schema, `<records>`.

Next, it is time to write out the data. We use the `.apply(...)` method of the DataFrame object to iterate through the records contained within. Its first parameter specifies the method to be applied to each record. By default, the `axis` parameter is set to `0`. This means that the method specified in the first parameter would be applied to each `column` of the DataFrame. By setting the parameter to `1`, we instruct the `.apply(...)` method that we want to apply the `xml_encode(...)` method specified in the first parameter to each `row`. We use the `xml_encode(...)` method to process each record from the `data` DataFrame:

```
def xml_encode(row):
    # first -- we output a record
    xmlItem = ['  <record>']
```

```
        # next -- for each field in the row we create a XML markup
        #          in a <field name=...>...</field> format
        for field in row.index:
            xmlItem \
                .append(
                    '    <var var_name="{0}">{1}</var>' \
                    .format(field, row[field])
                )


        # last -- this marks the end of the record
        xmlItem.append('  </record>')

        # return a string back to the calling method
        return '\n'.join(xmlItem)
```

The code creates a list of strings, `xmlItem`. The first element of the list is the `<record>` indicator and the last one will be `</record>`. Then, for each field in the row, we append values of each column for that record encapsulated within `<var var_name=`**<column_name>**`>`**<value>**`</var>`. The variables in bold indicate specific column names from the record (`<column_name>`) and corresponding value (`<value>`). Once all the fields of the record have been parsed, we create a long string by concatenating all the items of the `xmlItem` list using the `'\n'.join(...)` method. Each `<var>...</var>` tag is then separated by `\n`. The string is returned to the caller (`write_xml`). Each record is further concatenated in the `write_xml(...)` method and then output to the file. We finish with the closing tag, `</records>`.

Retrieving HTML pages with pandas

Although not as popular to store large datasets as previous formats, sometimes we find data in a table on a web page. These structures are normally enclosed within the `<table> </table>` HTML tags. This recipe will show you how to retrieve data from a web page.

Getting ready

In order to execute the following recipe, you need `pandas` and `re` modules available. The `re` module is a regular expressions module for Python and we will use it to clean up the column names. Also, the `read_html(...)` method of `pandas` requires `html5lib` to be present on your computer. If you use the Anaconda distribution of Python, you can do it by issuing the following command from your command line:

```
conda install html5lib
```

Otherwise, you can download the source from `https://github.com/html5lib/html5lib-python`, unzip it, and install the module manually:

cd html5lib-python-parser

python setup.py install

No other prerequisites are required.

How to do it...

`pandas` makes it very easy to access, retrieve, and parse HTML files. All this can be done in two lines. The `retrieve_html.py` file contains more code than that and we will discuss it in the next section:

```
# url to retrieve
url = 'https://en.wikipedia.org/wiki/' + \
      'List_of_airports_by_IATA_code:_A'

# extract the data from the HTML
url_read = pd.read_html(url, header = 0)[0]
```

How it works...

The `read_html(...)` method of `pandas` parses the DOM of the HTML file and retrieves the data from all the tables. It accepts a URL, file, or raw string with HTML tags as the first parameter. In our example, we also specified `header = 0` to extract the header from the table. The `read_html(...)` method returns a list of DataFrame objects, one for each table in the HTML file. The list of airports page from Wikipedia contains only one table so we only retrieve the first element from the returned list of DataFrames. That's it! The list of airports is already in the `url_read` object.

However, there are two issues with the data retrieved: column names contain whitespaces and separator rows are in the data. As the names can contain all variety of whitespace characters (space, tabulator, and so on), we use the `re` module:

```
import re

# regular expression to find any white spaces in a string
space = re.compile(r'\s+')

def fix_string_spaces(columnsToFix):
    '''
        Converts the spaces in the column name to underscore
    '''
    tempColumnNames = [] # list to hold fixed column names
```

```
# loop through all the columns
for item in columnsToFix:
    # if space is found
    if space.search(item):
        # fix and append to the list
        tempColumnNames \
            .append('_'.join(space.split(item)))
    else:
        # else append the original column name
        tempColumnNames.append(item)

return tempColumnNames
```

First, we compile the regular expression that attempts to find at least one space in a word.

 It goes beyond the scope of this book to discuss regular expressions in detail. A good compendium of knowledge on this topic can be found at `https://www.packtpub.com/application-development/mastering-python-regular-expressions` or in the `re` module documentation found at `https://docs.python.org/3/library/re.html`.

The method then loops through all the columns and, if it finds a space in the (`space.search(...)`) name, it then splits the column name (`space.split(...)`) into a list. The list is then concatenated using `'_'` as a separator. If, however, the column name contains no spaces, the original name is appended to the list of column names. To alter the column names in the DataFrame, we use the following code:

```
url_read.columns = fix_string_spaces(url_read.columns)
```

If you look at Wikipedia's list of airports table, you can see that it contains separator rows to group IATA names according to the first two letters of the code. All the other columns in the tables are missing. To deal with this issue, we can use DataFrame's `.dropna(...)` method.

 pandas has a couple of methods to deal with NaN (Not a Number) observations. In the *Imputing missing observations* recipe, we introduce the `.fillna(...)` method.

The .dropna(...) method drops rows (or columns if we want to) that contain at least one missing observation. However tempting it may be to just use .dropna(...) without any parameters, you would also drop all the legit rows that miss the **Daylight Saving Time** (**DST**) or ICAO code. We can, however, specify a threshold. A very rough look at the data leads to a conclusion that some legit records can contain up to two missing variables. The inplace=True parameter removes the data from the original DataFrame instead of creating a copy of the original one and returning a trimmed DataFrame; the default is inplace=False:

```
url_read.dropna(thresh=2, inplace=True)
```

Once we remove some rows, the DataFrame index will have holes. We can recreate it using the following code:

```
url_read.index = range(0,len(url_read))
```

To print out the top 10 IATA codes with corresponding airport names, we can use the following code:

```
print(url_read.head(10)[['IATA', 'Airport_name']])
```

If you want to retrieve more than one column, you put that in the form of a list; in our case, this was ['IATA', 'Airport_name']. The same results can be attained with the following code:

```
print(url_read[0:10][['IATA', 'Airport_name']])
```

Storing and retrieving from a relational database

The relational database model was invented in 1970 at IBM. Since then, it reigned the field of data analytics and storage for decades. The model is still widely used but has been losing the field to more and more popular solutions such as Hadoop. Nevertheless, the demise of relational databases is nowhere near as it is still a tool of choice for many applications.

Getting ready

In order to execute the following recipe, you need pandas and SQLAlchemy modules installed. The SQLAlchemy is a module that abstracts the database interactions between Python scripts and a number of relational databases; effectively, you do not have to remember the specifics of each database's syntax as SQLAlchemy will handle that for you.

If you are using the Anaconda distribution, issue the following command:

```
conda install sqlalchemy
```

Refer to your distribution of Python to check how to install new modules. Alternatively, check the previous recipe for instructions on how to install modules manually.

In addition, you might need to install the `psycopg2` module. This can be accomplished with the following command:

conda install psycopg2

If, however, you are not using Anaconda, the `psycopg2` can be found at `http://initd.org/psycopg/`. Follow the installation instructions found on the website.

To execute the script in this recipe, you will also need a PostgreSQL database running locally. Go to `http://www.postgresql.org/download/` and follow the installation instructions for your operating system contained therein. We assume that you have your PostgreSQL database installed up and running before you proceed. We also assume that your database can be accessed at `localhost:5432`. On a Unix-like system, you can check the port used by the PostgreSQL database by issuing the following command in the terminal:

cat /etc/services | grep postgre

No other prerequisites are required.

How to do it...

`pandas` works hand in hand with `SQLAlchemy` to make accessing and storing/retrieving data in/from many databases very easy. Reading the data and storing it in the database can be achieved with the following script (the `store_postgresql.py` file):

```python
import pandas as pd
import sqlalchemy as sa

# name of the CSV file to read from
r_filenameCSV = '../../Data/Chapter01/realEstate_trans.csv'

# database credentials
usr  = 'drabast'
pswd = 'pAck7!B0ok'

# create the connection to the database
engine = sa.create_engine(
    'postgresql://{0}:{1}@localhost:5432/{0}' \
    .format(usr, pswd)
)

# read the data
```

```
csv_read = pd.read_csv(r_filenameCSV)

# transform sale_date to a datetime object
csv_read['sale_date'] = pd.to_datetime(csv_read['sale_date'])

# store the data in the database
csv_read.to_sql('real_estate', engine, if_exists='replace')
```

How it works...

First, we import all the necessary modules: `pandas` and `SQLAlchemy`. We also store the name of the CSV file we will be reading the data from in a variable, and we specify the credentials to be used to connect to our PostgreSQL database. Then, using SQLAchemy's `create_engine(...)` method, we create an object that allows us to access the PostgreSQL database. The connection string specific syntax can be broken down as follows:

```
sa.create_engine('postgresql://<user_name>:<password>@<server>:<port>
/<database>')
```

Here, `<user_name>` is the username allowed to log in to `<database>` using `<password>`. The user needs (at a minimum) CREATE, DROP, INSERT, and SELECT privileges for the specified database. The `<server>` tag can be expressed as an IP address of the server running the PostgreSQL database or (as in our case) the name of the server (`localhost`). The `<port>` specifies the server port the database listens on.

Next, we read in the data from a CSV file and convert the `sale_date` column to a `datetime` object so that we can store the data in a date format in the database. The `read_csv(...)` method normally tries to infer the proper format for the data read from a file but it can get really tricky with dates. Here, we explicitly specify the date format for the `sale_date` column.

The last line of the script stores the information in the database. The `to_sql(...)` method specifies the name of the table (`'real_estate'`) and connector (`engine`) to be used. The last parameter passed instructs the method to replace the table if it already exists in the database.

You can check whether the data has loaded properly using the following command:

```
query = 'SELECT * FROM real_estate LIMIT 10'
top10 = pd.read_sql_query(query, engine)
print(top10)
```

We first specify a valid SQL query and then use the `read_sql_query(...)` method to execute it.

There's more...

The most popular database in the world is SQLite. SQLite databases can be found in phones, TV sets, cars, among others; it makes SQLite the most widespread database. SQLite is very lightweight and requires no server to run. It can either store the data on a disk or use the memory of your computer to temporarily keep the data. The latter can be used when speed is required but the data disappears as soon as your script finishes.

With SQLAlchemy, it is also extremely easy to talk to the SQLite database. The only change required in the preceding example is how we construct the engine (the store_SQLite.py file):

```
# name of the SQLite database
rw_filenameSQLite = '../../Data/Chapter01/realEstate_trans.db'

# create the connection to the database
engine = sa.create_engine(
    'sqlite:///{0}'.format(rw_filenameSQLite)
)
```

As you can see, as the SQLite databases are serverless, the only required parameter is where to store the database file itself.

If, instead of storing the database in the file, you would like to keep your data in the computer's memory, use sqlite:// as the connection string.

Note the three slashes in the path; this is to help the innards of the create_engine(...) method. At the most general level, the connection string follows the following pattern:

<database_type>://<server_information>/<database>

As SQLite databases do not require any server, <server_information> is empty and, hence, three slashes.

See also

I highly recommend checking out the documentation for SQLAlchemy as it is a very powerful middleman between your code and various databases; the documentation can be found at http://docs.sqlalchemy.org/en/rel_1_0/index.html.

Storing and retrieving from MongoDB

MongoDB has become one of the most popular, so-called, NoSQL databases that are there. It is highly scalable, has a very flexible data model and highly expressive query language. MongoDB is highly popular for storing unstructured data, especially for web-based apps: MEAN.js combines MongoDB, Express.js, Angular.js, and Node.js in one fullstack framework.

Getting ready

You need to have pandas and PyMongo modules installed. If you do not have PyMongo, you can either use conda to install the missing package or go to http://api.mongodb.org/python/current/installation.html#installing-from-source and follow the steps listed there.

Also, you need the MongoDB database up and running on your computer. You can download MongoDB from https://www.mongodb.org/downloads and then follow the instructions listed at http://docs.mongodb.org/manual/installation/ for your operating system.

No other prerequisites are required.

How to do it...

Accessing and storing documents in MongoDB is almost effortless. We assume that the data to store in the database are accessible to us in a pandas' DataFrame object csv_read (the store_mongodb.py file):

```
# connect to the MongoDB database
client = pymongo.MongoClient()

# and select packt database
db = client['packt']

# then connect to real_estate collection
real_estate = db['real_estate']

# and then insert the data
real_estate.insert(csv_read.to_dict(orient='records'))
```

How it works...

We first create a connection to MongoDB. As we are not passing any arguments to the `MongoClient(...)` method, PyMongo will connect to the default MongoDB database running on `localhost` and listening on port `27017`. If you would like to connect to a different database, you can override these defaults; assuming that the database you want to connect to runs on a server with IP `10.92.1.12` and listens on a port `1234`, you can use the following code:

```
client = pymongo.MongoClient(host='10.92.1.12', port=1234)
```

 A full list of the `MongoClient(...)` method's available parameters with explanations is available at `http://api.mongodb.org/python/current/api/pymongo/mongo_client.html`.

The `MongoClient(...)` method automatically connects to the MongoDB database so there is no need to do it manually. Once connected, we connect to the internal database object we are after (in our case, we called it `packt`) and we store it in the `db` object. Depending on whether the `real_estate` collection exists or not in the database, the next line of our script either opens the collection or creates a new one, respectively. We then get access to this collection through the `real_estate` object.

 In MongoDB, collections are equivalents of tables. Each collection consists of documents, equivalents to records from the relational database world.

Then, and only then, we can start storing our documents in the collection. We use the `insert(...)` method to load the data to the collection.

 We use version 2.8 of PyMongo. In the newest (at the time of writing this book) stable version 3.0, the `insert(...)` method was deprecated in favor of `insert_many(...)`. Refer to the documentation of your version of PyMongo at `http://api.mongodb.org/python/2.8/api/index.html` or `http://api.mongodb.org/python/3.0/api/index.html`.

The `insert(...)` method accepts a list of dictionaries. We use DataFrame's `to_dict(...)` method that does all the heavy lifting for us in transforming the data into a list of dictionaries. The `orient='records'` option instructs the method to return a list where each element is a dictionary with column names as keys and corresponding values. Each element of the resulting list will form a document in the collection.

The `insert(...)` method appends the data by default. If this is not desired, you can use the following code to remove all the documents from the collection before inserting into it:

```
# if there are any documents stored already -- remove them
if real_estate.count() > 0:
    real_estate.remove()
```

The `count()` method calculates the total number of documents in the collection—if there are any, the `remove()` method deletes all the documents.

Next, we want to print top 10 sales (by record ID) from ZIP codes `95841` and `95842`. We can do it easily by querying our collection:

```
sales = real_estate.find({'zip': {'$in': [95841, 95842]}})
for sale in sales.sort('_id').limit(10):
    print(sale)
```

First, we extract all the sales from these ZIP codes. In our data, the field name we are after is `zip`. The `$in` operator works in the same way as the ANSI SQL expression `IN (...)` so we can specify a list of all the ZIP codes that we want to retrieve from the database. The `find(...)` method returns a `cursor` object; the cursor object is a generator. So, we use it to iterate through all the sales and print out each individual sale to the screen.

See also

I suggest referring to the documentation of MongoDB at `http://docs.mongodb.org/v2.6/` and `PyMongo` for more details.

Opening and transforming data with OpenRefine

OpenRefine originated as GoogleRefine. Google later open sourced the code. It is a great tool to sift through the data quickly, clean it, remove duplicate rows, analyze distributions or trends over time, and more.

In this and the following recipes, we will deal with the `realEstate_trans_dirty.csv` file that is located in the `Data/Chapter1` folder. The file has several issues that, over the course of the following recipes, we will see how to resolve.

First, when read from a text file, OpenRefine defaults the types of data to text; we will deal with data type transformations in this recipe. Otherwise, we will not be able to use facets to explore the numerical columns. Second, there are duplicates in the dataset (we will deal with them in the *Remove duplicates* recipe). Third, the `city_state_zip` column, as the name suggests, is an amalgam of city, state, and zip. We prefer keeping these separate, and in the *Using regular expressions and GREL to clean up data* recipe, we will see how to extract such information. There is also some missing information about the sale price—we will impute the sale prices in the *Imputing missing observations* recipe.

Getting ready

To run through these examples, you need OpenRefine installed and running on your computer. You can download OpenRefine from `http://openrefine.org/download.html`. The installation instructions can be found at `https://github.com/OpenRefine/OpenRefine/wiki/Installation-Instructions`.

OpenRefine runs in a browser so you need an Internet browser installed on your computer. I tested it in Chrome and Safari and found no issues.

> The Mac OS X Yosemite comes with Java 8 installed by default. OpenRefine does not support it. You need to install Java 6 or 7—see `https://support.apple.com/kb/DL1572?locale=en_US`.
>
> However, even after installing legacy versions of Java, I still experienced some issues with version 2.5 of OpenRefine on Mac OS X Yosemite and El Capitan. Using the beta version (2.6), even though it is still in development, worked fine.

No other prerequisites are required.

How to do it...

First, you need to start OpenRefine, open your browser, and type `http://localhost:3333`. A window similar to the following screenshot should open:

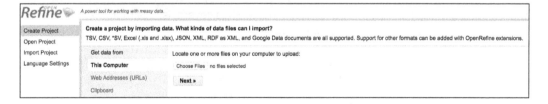

The first thing you want to do is create a project. Click on **Choose files**, navigate to **Data/Chapter1**, and select **realEstate_trans_dirty.csv**. Click **OK**, then **Next**, and **Create Project**. After the data opens, you should see something similar to this:

Note that the **beds**, **baths**, **sq__ft**, **price**, **latitude**, and **longitude** data is treated as text and so is **sale_date**. While converting the former is easy, the format of **sale_date** is not as easy to play with in OpenRefine:

If the text data was in a format resembling, for example, **2008-05-21**, we could just use the **Google Refine Expression Language** (**GREL**) method `.toDate()` and OpenRefine would convert the dates for us. In our case, we need to use some trickery to convert the dates properly. First, we select a **Transform** option, as shown in the following screenshot:

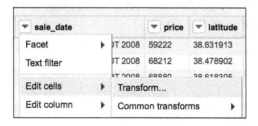

Then, in the window that opens, we will use GREL to convert the dates as follows:

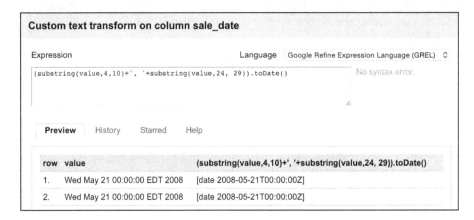

The `value` variable here represents the value of each cell in the selected column (`sale_date`). The first part of the expression extracts the month and day from the value, that is, we get in return `May 21` by specifying that we want to retrieve a substring starting at the fourth character and finishing at the tenth character. The second `substring(...)` method extracts the year from the string. We separate the two by a comma using the `...+',` `'+...` expression. The resulting value will result in the `May 21, 2008` string pattern. Now OpenRefine can deal with this easily. Thus, we wrap our two `substring` methods inside parentheses and use the `.toDate()` method to convert the date properly. The **Preview** tab in the right column shows you the effect of our expression.

See also

A very good introduction and deep dives into the various aspects of OpenRefine can be found in the *Using OpenRefine* book by *Ruben Verborgh* and *Max De Wilde* at `https://www.packtpub.com/big-data-and-business-intelligence/using-openrefine`.

Exploring the data with Open Refine

Understanding your data is the first step to build successful models. Without intimate knowledge of your data, you might build a model that performs beautifully in the lab but fails gravely in production. Exploring the dataset is also a great way to see if there are any problems with the data contained within.

Getting ready

To follow this recipe, you need to have OpenRefine and virtually any Internet browser installed on your computer. See the *Opening and transforming data with OpenRefine* recipe's *Getting ready* subsection to see how to install OpenRefine.

We assume that you followed the previous recipe so your data is already loaded to OpenRefine and the data types are now representative of what the columns hold. No other prerequisites are required.

How to do it...

Exploring data in OpenRefine is easy with Facets. An OpenRefine Facet can be understood as a filter: it allows you to quickly either select certain rows or explore the data in a more straightforward way. A facet can be created for each column—just click on the down-pointing arrow next to the column and, from the menu, select the **Facet** group.

There are four basic types of facet in OpenRefine: text, numeric, timeline, and scatterplot.

You can create your own custom facets or also use some more sophisticated ones from the OpenRefine arsenal such as word or text lengths facets (among others).

The text facet allows you to get a sense of the distribution of text columns from your dataset quickly. For example, we can see which city in our dataset had the most sales between May 15, 2008 and May 21, 2008. As expected, since we analyze data from the Sacramento area, the city tops the list, followed by Elk Grove, Lincoln, and Roseville, as shown in the following screenshot:

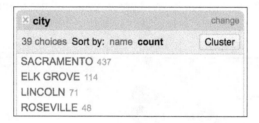

This gives you a very easy and straightforward insight on whether the data makes sense or not; you can readily determine whether the provided data is what it was supposed to be.

The numeric facet allows you to glimpse the distribution of your numeric data. We can, for instance, check the distribution of prices in our dataset, as shown in the following screenshot:

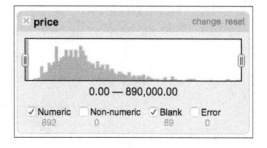

The distribution of the prices roughly follows what we would expect: left (positive) skewed distribution of sales prices makes sense, as one would expect less sales at the far right end of the spectrum, that is, people having money and willingness to purchase a 10-bedroom villa.

This facet reveals one of the flaws of our dataset: there are 89 missing observations in the price column. We will deal with these later in the book, in the *Imputing missing observations* recipe.

It is also good to check whether there are any blanks in the timeline of sales, as we were told that we would get seven days of data (May 15 to May 21, 2008):

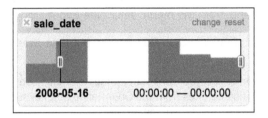

Our data indeed spans seven days but we see two days with no sales. A quick check of the calendar reveals that May 17 and May 18 was a weekend so there are no issues here. The timeline facet allows you to filter the data using the sliders on each side; here, we filtered observations from May 16, 2008 onward.

The scatterplot facet lets you analyze interactions between all the numerical variables in the dataset:

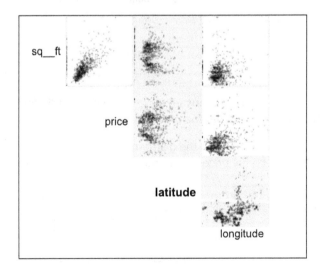

By clicking at the particular row and column, you can analyze the interactions in greater detail:

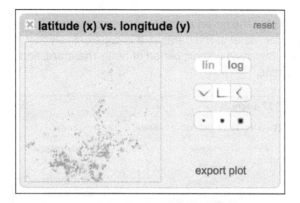

Removing duplicates

We can safely assume that all the data that lands on our desks is dirty (until proven otherwise). It is a good habit to check whether everything with our data is in order. The first thing I always check for is the duplication of rows.

Getting ready

To follow this recipe, you need to have OpenRefine and virtually any Internet browser installed on your computer.

We assume that you followed the previous recipes and your data is already loaded to OpenRefine and the data types are now representative of what the columns hold. No other prerequisites are required.

How to do it...

First, we assume that within the seven days of property sales, a row is a duplicate if the same address appears twice (or more) in the dataset. It is quite unlikely that the same house is sold twice (or more times) within such a short period of time. Therefore, first, we **Blank down** the observations if they repeat:

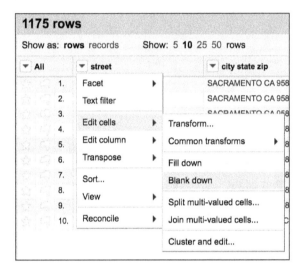

This effects in keeping only the first occurrence of a certain set of observations and blanking the rest (see the fourth row in the following screenshot):

 The **Fill down** option has the opposite effect—it would fill in the blanks with the values from the row above unless the cell is not blank.

We can now create a Facet by blank that would allow us to quickly select the blanked rows:

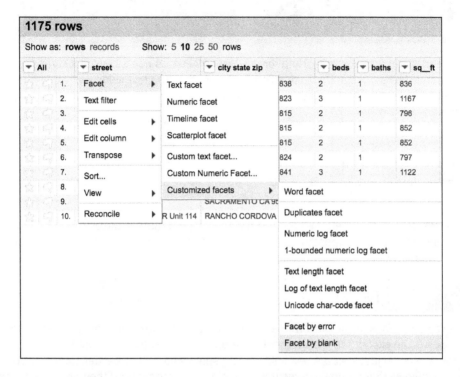

Creating such a facet allows us to quickly select all the rows that are blank and remove them from the dataset:

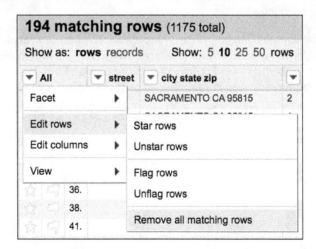

Our dataset now has no duplicate records.

Using regular expressions and GREL to clean up data

When cleaning up and preparing data for use, we sometimes need to extract some information from text fields. Occasionally, we can just split the text fields using delimiters. However, when a pattern of data does not allow us to simply split the text, we need to revert to regular expressions.

Getting ready

To follow this recipe, you need to have OpenRefine and virtually any Internet browser installed on your computer.

We assume that you followed the previous recipes and your data is already loaded to OpenRefine and the data types are now representative of what the columns hold. No other prerequisites are required.

How to do it...

First, let's have a look at the pattern that occurs in our **city_state_zip** column. As the name suggests, we can expect the first element to be the city followed by state and then a 5-digit ZIP code. We could just split the text field using a space character as a delimiter and be done with it. It would work for many records (for example, Sacramento) and they would be parsed properly into city, state, and ZIP. There is one problem with this approach—some locations consist of two or three words (for example, Elk Grove). Hence, we need a slightly different approach to extract such information.

This is where regular expressions play an invaluable role. You can use regular expressions in OpenRefine to transform the data. We will now split **city_state_zip** into three columns: city, state, and zip. Click on the downward button next to the name of the column and, from the menu, select **Edit column** and **Add column** based on this column. A window should appear, as shown in the following screenshot:

As before, the value represents the value of each cell. The `.match(...)` method is applied to the cell's value. It takes a regular expression as its parameter and returns a list of values matched given the expressed pattern. The regular expression is encapsulated between `/.../`. Let's break the regular expression down step by step.

We know the pattern of the **city_state_zip** column: first is the name of the city (can be more than one word), followed by a two-character state acronym, and ending with a 5-digit ZIP code. The regular expression to match such a pattern will be as follows:

```
(.*) (..) (\d{5})
```

It is easier to read this expression starting from the end. So, reading from the right, first we extract the ZIP code using `(\d{5})`. The `\d` indicates any digit (and is equivalent to stating `([0-9]{5})`) and `{5}` selects five digits from the back of the string. Next, we have `(..)¬`. This expression extracts the two-character acronym of the state identified by two dots `(..)`. Note that we used ¬ in place of a space character just for readability purposes. This expression extracts only two characters and a space from the string—no less, no more. The last (reading from the right) is `(.*)` that can be understood as: extract all the characters (if any) that will not be matched by the other two expressions.

In entirety, the expression can be translated into English as follows: extract a string (even if empty) until a two-character acronym of the state is encountered (preceded by a space character) followed by a space and five digits indicating the ZIP code.

The `.match(...)` method generates a list. In our case, we will get back a list of three elements. To extract city, we select the first element from that list `[0]`. To select state and ZIP, we will repeat the same steps but select `[1]` and `[2]` respectively.

Now that we're done with splitting the **city_state_zip** column, we can export the project to a file. In the top right corner of the tool, you will find the **Export** button; select **Comma-separated value**. This will download the file automatically to your `Downloads` folder.

See also

I highly recommend reading the *Mastering Python Regular Expressions* book by *Felix Lopez* and *Victor Romero* available at `https://www.packtpub.com/application-development/mastering-python-regular-expressions`.

Imputing missing observations

Collecting data is messy. Research data collection instruments fail, humans do not want to answer some questions in a questionnaire, or files might get corrupted; these are but a sample of reasons why a dataset might have missing observations. If we want to use the dataset, we have a couple of choices: remove the missing observations altogether or replace them with some value.

Getting ready

To execute this recipe, you will need the `pandas` module.

No other prerequisites are required.

How to do it...

Once again, we assume that the reader followed the earlier recipes and the `csv_read` DataFrame is already accessible to us. To impute missing observations, all you need to do is add this snippet to your code (the `data_imput.py` file):

```
# impute mean in place of NaNs
csv_read['price_mean'] = csv_read['price'] \
    .fillna(
        csv_read.groupby('zip')['price'].transform('mean')
    )
```

How it works...

The `pandas'` `.fillna(...)` method does all the heavy lifting for us. It is a DataFrame method that takes the value to be imputed as its only required parameter.

> Consult the `pandas` documentation of `.fillna(...)` to see other parameters that can be passed to the method. The documentation can be found at `http://pandas.pydata.org/pandas-docs/stable/generated/pandas.DataFrame.fillna.html`.

In our approach, we assumed that each ZIP code might have different price averages. This is why we first grouped the observations using the `.groupby(...)` method. It stands to reason that the prices of houses would also heavily depend on the number of rooms in a given house; if our dataset had more observations, we would have added the `beds` variable as well.

The `.groupby(...)` method returns a `GroupBy` object. The `.transform(...)` method of the `GroupBy` object effectively replaces all the observations within ZIP code groups with a specified value, in our case, the mean for each ZIP code.

The `.fillna(...)` method now simply replaces the missing observations with the mean of the ZIP code.

There's more...

Imputing the mean is not the only way to fill in the blanks. It returns a reasonable value only if the distribution of prices is symmetrical and without many outliers; if the distribution is skewed, the average is biased. A better metric of central tendency is the median. It takes one simple change in the way we presented earlier:

```
# impute median in place of NaNs
csv_read['price_median'] = csv_read['price'] \
    .fillna(
        csv_read.groupby('zip')['price'].transform('median')
    )
```

Normalizing and standardizing the features

We normalize (or standardize) data for computational efficiency and so we do not exceed the computer's limits. It is also advised to do so if we want to explore relationships between variables in a model.

Computers have limits: there is an upper bound to how big an integer value can be (although, on 64-bit machines, this is, for now, no longer an issue) and how good a precision can be for floating-point values.

Normalization transforms all the observations so that all their values fall between 0 and 1 (inclusive). Standardization shifts the distribution so that the mean of the resultant values is 0 and standard deviation equals 1.

Getting ready

To execute this recipe, you will need the `pandas` module.

No other prerequisites are required.

How to do it...

To perform normalization and standardization, we define two helper functions (the `data_standardize.py` file):

```
def normalize(col):
    '''
        Normalize column
    '''
    return (col - col.min()) / (col.max() - col.min())

def standardize(col):
    '''
        Standardize column
    '''
    return (col - col.mean()) / col.std()
```

How it works...

To normalize a set of observations, that is, to make each and every single one of them to be between 0 and 1, we subtract the minimum value from each observation and divide it by the range of the sample. The range in statistics is defined as a difference between the maximum and minimum value in the sample. Our `normalize(...)` method does exactly as described previously: it takes a set of values, subtracts the minimum from each observation, and divides it by the range.

Standardization works in a similar way: it subtracts the mean from each observation and divides the result by the standard deviation of the sample. This way, the resulting sample has a mean equal to 0 and standard deviation equal to 1. Our `standardize(...)` method performs these steps for us:

```
csv_read['n_price_mean'] = normalize(csv_read['price_mean'])
csv_read['s_price_mean'] = standardize(csv_read['price_mean'])
```

Binning the observations

Binning the observations comes in handy when we want to check the shape of the distribution visually or we want to transform the data into an ordinal form.

Getting ready

To execute this recipe, you will need the `pandas` and `NumPy` modules.

No other prerequisites are required.

How to do it...

To bin your observations (as in a histogram), you can use the following code (data_binning.py file):

```
# create bins for the price that are based on the
# linearly spaced range of the price values
bins = np.linspace(
    csv_read['price_mean'].min(),
    csv_read['price_mean'].max(),
    6
)

# and apply the bins to the data
csv_read['b_price'] = np.digitize(
    csv_read['price_mean'],
    bins
)
```

How it works...

First, we create bins. For our price (with the mean imputed in place of missing observations), we create six bins, evenly spread between the minimum and maximum values for the price. The `.linspace(...)` method does exactly this: it creates a `NumPy` array with six elements, each greater than the preceding one by the same value. For example, a `.linspace(0,6,6)` command would generate an array, `[0., 1.2, 2.4, 3.6, 4.8, 6.]`.

NumPy is a powerful numerical library for linear algebra. It can easily handle large arrays and matrices and offers a plethora of supplemental functions to operate on such data. For more information, visit http://www.numpy.org.

The .digitize(...) method returns, for each value in the specified column, the index of the bin that the value belongs to. The first parameter is the column to be binned and the second one is the array with bins.

To count the records within each bin, we use the .value_counts() method of DataFrame, counts_b = csv_read['b_price'].value_counts().

There's more...

Sometimes, instead of having evenly-spaced values, we would like to have equal counts in each bucket. To attain such a goal, we can use **quantiles**.

Quantiles are closely related to percentiles. The difference is percentiles return values at a given sample percentage, while quantiles return values at the sample fraction. For more information, visit https://www.stat.auckland.ac.nz/~ihaka/787/lectures-quantiles-handouts.pdf.

What we want to achieve is splitting our column into deciles, that is, 10 bins of (more or less) equal size. To do this, we can use the following code (you can easily spot the similarities with the previous approach):

```
# create bins based on deciles
decile = csv_read['price_mean'].quantile(np.linspace(0,1,11))

# and apply the decile bins to the data
csv_read['p_price'] = np.digitize(
    csv_read['price_mean'],
    decile
)
```

The .quantile(...) method can either take one number (between 0 and 1) indicating the percentile to return (for example, 0.5 being the median and 0.25 and 0.75 being lower and upper quartiles). However, it can also return an array of values corresponding to the percentiles passed as a list to the method. The .linspace(0,1,11) command will produce the following array:

```
[ 0., 0.1, 0.2, 0.3, 0.4, 0.5, 0.6, 0.7, 0.8, 0.9, 1. ]
```

So, the .quantile(...) method will return a list starting with a minimum and followed by all the deciles up to the maximum for the price_mean column.

Encoding categorical variables

The final step on the road to prepare the data for the exploratory phase is to bin categorical variables. Some software packages do this behind the scenes, but it is good to understand when and how to do it.

Any statistical model can accept only numerical data. Categorical data (sometimes can be expressed as digits depending on the context) cannot be used in a model straightaway. To use them, we encode them, that is, give them a unique numerical code. This is to explain when. As for how—you can use the following recipe.

Getting ready

To execute this recipe, you will need the `pandas` module.

No other prerequisites are required.

How to do it...

Once again, `pandas` already has a method that does all of this for us (the `data_dummy_code.py` file):

```
# dummy code the column with the type of the property
csv_read = pd.get_dummies(
    csv_read,
    prefix='d',
    columns=['type']
)
```

How it works...

The `.get_dummies(...)` method converts categorical variables into dummy variables. For example, consider a variable with three different levels:

```
1   One
2   Two
3   Three
```

We will need three columns to code it:

```
1   One    1   0   0
2   Two    0   1   0
3   Three  0   0   1
```

Sometimes, we can get away with using only two additional columns. However, we can use this trick only if one of the levels is, effectively, null:

```
1   One   1   0
2   Two   0   1
3   Zero  0   0
```

The first parameter to the `.get_dummies(...)` method is the DataFrame. The `columns` parameter specifies the column (or columns, as we can also pass a list) in the DataFrame to the dummy code. Specifying the prefix, we instruct the method that the names of the new columns generated should have the `d_` prefix; in our example, the generated dummy-coded columns will have `d_Condo` names (as an example). The underscore `_` character is default but can also be altered by specifying the `prefix_sep` parameter.

For a full list of parameters to the `.get_dummies(...)` method, see `http://pandas.pydata.org/pandas-docs/stable/generated/pandas.get_dummies.html`.

2
Exploring the Data

In this chapter, we will cover various techniques that will allow you to understand your data better and explore relationships between features. You will learn the following recipes:

- ▸ Producing descriptive statistics
- ▸ Exploring correlations between features
- ▸ Visualizing interactions between features
- ▸ Producing histograms
- ▸ Creating multivariate charts
- ▸ Sampling the data
- ▸ Splitting the dataset into training, cross-validation, and testing

Introduction

In the following recipes, we will use Python and D3.js to build our understanding of the data. We will analyze the distributions of all the variables, investigate the correlations between features, and visualize the interactions between them. You will learn how to generate histograms and present three-dimensional data on a two-dimensional chart. Finally, you will learn how to produce stratified samples and split your dataset into testing and training subsets.

Producing descriptive statistics

To fully understand the distribution of any random variable, we need to know its mean and standard deviation, minimum and maximum values, median, mode, first and third quartiles, skewness, and kurtosis.

Sometimes, it is good to perform statistical testing to confirm (or disprove) whether our data follows a specific distribution. This, however, is beyond the scope of this book.

Getting ready

To execute this recipe, all you need is `pandas`. No other prerequisites are required.

How to do it...

Here is a piece of code that can quickly give you a basic understanding of your data. We assume that our data was read from a CSV file and stored in the `csv_read` variable (the `data_describe.py` file):

```
# calculate the descriptives: count, mean, std,
# min, 25%, 50%, 75%, max
# for a subset of columns
csv_desc = csv_read[
    [
        'beds','baths','sq__ft','price','s_price',
        'n_price','s_sq__ft','n_sq__ft','b_price',
        'p_price','d_Condo','d_Multi-Family',
        'd_Residential','d_Unkown'
    ]
].describe().transpose()

# and add skewness, mode and kurtosis
csv_desc['skew'] = csv_read.skew(numeric_only=True)
csv_desc['mode'] = \
    csv_read.mode(numeric_only=True).transpose()
csv_desc['kurtosis'] = csv_read.kurt(numeric_only=True)
```

How it works...

`pandas` has a very useful `.describe()` method that does most of the work for us. This method produces most of the descriptive statistics that we are after; the output looks as follows (abbreviated for legibility purposes):

```
             beds
count   981.000000
mean      2.914373
std       1.306502
min       0.000000
25%       2.000000
50%       3.000000
75%       4.000000
max       8.000000
```

The index of the DataFrame shows the name of the descriptive statistics, and each column represents a specific variable from our dataset. However, we are still missing the skewness, kurtosis, and mode. In order to make it easier for us to add to our `csv_desc` variable, we `.transpose()` the output of the `.describe()` method so that the names of the variables are stored in the index and each column represents the descriptive variable.

> You can also calculate the mean, standard deviation, or any other descriptive statistics by hand (if you do not need all of them). Check `http://pandas.pydata.org/pandas-docs/stable/api.html#api-dataframe-stats` for the full list of applicable methods.

Having the basic descriptive data in place, we need to add the missing ones. You need to be careful as the `.skew(...)` and `.kurt(...)` methods return the data in a similar format while `.mode(...)` does not; the returned data from the `.mode(...)` method needs to be transposed as it is stored in the same format as the output of `.describe(...)`.

There's more...

The descriptive statistics can also be calculated using `SciPy` and `NumPy`. Things are a bit less intuitive than `pandas` (file `data_describe_alternative.py`).

First, we load the modules.

```
import scipy.stats as st
import numpy as np
```

Once this is done, we read in the data from a CSV file:

```
r_filenameCSV = '../../Data/Chapter02/' + \
    'realEstate_trans_full.csv'

csv_read = np.genfromtxt(
    r_filenameCSV,
    delimiter=',',
    names=True,
    # only numeric columns
    usecols=[4,5,6,8,11,12,13,14,15,16,17,18,19,20]
)
```

The `.genfromtxt(...)` method takes the name of the file as its first (and only required) parameter. It is a good practice to specify the delimiter; in our case, this is `','` but can also be `\t`. The `names` parameter, when set to `True`, instructs the method that the names of the variables are stored in the first row. Finally, the `usecols` parameter specifies which columns from the file to actually store in the `csv_read` object.

Finally, we calculate the required statistics:

```
desc = st.describe([list(item) for item in csv_read])
```

The data that is created by the `.genfromtxt(...)` method is a list of tuples. The `.describe(...)` method accepts the data in a list form only, so before we can use it, we cast each tuple into a list (using list comprehension).

The output of the method is arguably less user-friendly:

```
endeavour:Chapter02 drabast$ python data_describe_alternative.py
DescribeResult(nobs=981, minmax=(array([ 0.        ,  0.        ,  0.        ,         nan,         nan,
               nan,         nan,         nan, 0.        , -1.54253538,
          1.        ,  1.        ,  0.        ,  0.        ]), array([ 8.00000000e+00,   5.00000000e+00,   5.82200000e+03,
               nan,         nan,         nan,
               nan,         nan, 1.00000000e+00,
          5.27789305e+00,  6.00000000e+00,   1.10000000e+01,
          1.00000000e+00,  1.00000000e+00])), mean=array([ 2.91437309e+00,   1.77879715e+00,   1.31672681e+03,
               nan,         nan,         nan,
               nan,         nan, 2.26164000e-01,
          2.71119066e-17,  1.84097859e+00,   5.51681957e+00,
          5.50458716e-02,  1.32517839e-02]), variance=array([ 1.70694626e+00,   8.01019368e-01,   7.28653560e+05,
               nan,         nan,         nan,
               nan,         nan, 2.14969422e-02,
          1.00000000e+00,  6.44074143e-01,   8.31323722e+00,
          5.20689010e-02,  1.30895172e-02]), skewness=array([ -7.94572093e-01,  -2.35612114e-01,   5.24629123e-01,
               nan,         nan,         nan,
               nan,         nan, 5.24629123e-01,
          5.24629123e-01,  1.14766678e+00,   5.28302734e-03,
          3.90191224e+00,  8.51322314e+00]), kurtosis=array([ 0.63188203,   0.35870044,   1.24352907,         nan,
               nan,         nan,         nan,         nan,
          1.24352907,   1.24352907,   2.53828402,  -1.22115733,
         13.22491909,  70.47496821]))
```

See also...

Check the documentation for `SciPy` to get a list of all statistical functions:

`http://docs.scipy.org/doc/scipy/reference/stats.html#statistical-functions`

Exploring correlations between features

A correlation coefficient between two variables measures the degree of relationship between them. If the coefficient is equal to `1`, we then say that the variables are perfectly correlated, if it is `-1`, then we conclude that the second variable is perfectly inversely correlated with the first one. The coefficient of `0` means that no measurable relationship exists between the two variables.

 We need to stress one fundamental truth here: one cannot conclude that simply if two variables are correlated, there exists a causal relationship between them. For more information, refer to the following website: `https://web.cn.edu/kwheeler/logic_causation.html`.

Getting ready

To execute this recipe, you need `pandas`. No other prerequisites are required.

How to do it...

We will be checking only correlations between the number of bedrooms that the apartment has, number of baths, the floor area, and price. Once again, we assume that the data is already in the `csv_read` object. Here's the code (the `data_correlations.py` file):

```
# calculate the correlations
coefficients = ['pearson', 'kendall', 'spearman']

csv_corr = {}

for coefficient in coefficients:
    csv_corr[coefficient] = csv_read \
        .corr(method=coefficient) \
        .transpose()
```

How it works...

`pandas` offers calculating three types of correlation: Pearson's product-moment correlation coefficient, Kendall's rank correlation coefficient, and Spearman's rank correlation coefficient, with the latter two being somewhat less sensitive to random variables with non-normal distributions.

We calculate all three and store the results in the `csv_corr` variable. The `.corr(...)` method is invoked on the DataFrame object, `csv_read`. The only required parameter is the method to be used. The results are presented as follows:

	coefficient	beds	baths	sq__ft	price
beds	pearson	1			
	kendall	1			
	spearman	1			
baths	pearson	0.84	1		
	kendall	0.71	1		
	spearman	0.76	1		
sq__ft	pearson	0.68	0.66	1	
	kendall	0.59	0.58	1	
	spearman	0.68	0.66	1	

	coefficient	beds	baths	sq__ft	price
price	pearson	0.36	0.42	0.36	1
	kendall	0.33	0.37	0.36	1
	spearman	0.41	0.44	0.43	1

See also...

You can also calculate the Pearson's coefficient of correlation using `NumPy`:

`http://docs.scipy.org/doc/numpy/reference/generated/numpy.corrcoef.html`.

Visualizing the interactions between features

D3.js is a powerful framework created by Mike Bostock to visualize data. It allows you to interactively manipulate data using HTML, SVG, and CSS. In this recipe, we will explore whether a relationship exists between the price of a property and the floor area.

Getting ready

To execute this recipe, you will need `pandas` and `SQLAlchemy` to prepare the data. For the visualization, all you need is the D3.js code (available in the GitHub repository for the book in the `/Data/Chapter02/d3 folder`). Some familiarity with HTML and JavaScript is required.

How to do it...

The code for this recipe comes in two parts: the data preparation (Python) and data visualization (HTML and D3.js).

The data preparation part is simple, and by now, you should be able to do it yourself. You can also refer to the *Storing and retrieving from a relational database* recipe from *Chapter 1, Preparing the Data*. We extract the data from the PostgreSQL database using `SQLAlchemy`. The query for our examples is as follows (the `data_interaction.py` file):

```
query = '''SELECT sq__ft,
                  price / 1000 AS price
           FROM real_estate
           WHERE sq__ft > 0
           AND beds BETWEEN 2 AND 4'''
data = pd.read_sql_query(query, engine)
```

We store the retrieved data in the `/Data/Chapter02/realEstate_d3.csv` file.

To visualize the data, we start with a clean HTML file.

 We will not be covering the basics of web development in this book as numerous online resources exist on how to prepare HTML documents. A good place to start is `http://www.w3schools.com/`.

To make the framework available to us, we need to import it. Here, we provide a pseudo-code on how to go about creating a scatterplot with D3.js. We will explain the code step by step in the next section:

```
<script src='d3.v3.min.js' charset='utf-8'>

// Append an SVG object to the HTML body

// Read in the dataset

// Find the maximum price and area to define scales of axes

// Define the scales

// Define the axes

// Draw dots on the chart

// Append X axis to chart

// Append Y axis to chart

// Append axis labels
```

How it works...

First, we append a **Scalable Vector Graphics** (**SVG**) object to the HTML **Document Object Model** (**DOM**):

```
var width   = 800;
var height  = 600;
var spacing = 60;

// Append an SVG object to the HTML body
var chart = d3.select('body')
    .append('svg')
    .attr('width',  width  + spacing)
    .attr('height', height + spacing)
;
```

Exploring the Data

We are using D3.js to select the body object from the DOM and append an SVG object to it, specifying its attributes, width, and height. What you see here is a chaining of methods and parameters; such a chain reads from left to right. It is equivalent to the following (somewhat less readable) code:

```
var body = d3.select('body');
var chart = body.append('svg');
chart.attr('width', width + spacing);
chart.attr('height', height + spacing);
```

Now that we have an SVG object appended to the DOM, it's time to read in the data. This can be achieved with the following code:

```
d3.csv('http://bit.ly/1LGB1rA', function(d) {
        draw(d) });
```

We use the method provided by D3.js to read in the CSV. The first parameter to the `.csv(...)` method specifies the dataset; in our case, we are reading the file that we stored in the GitHub repository for this book. (Go ahead and check it!) The second parameter is a callback to a function that will ingest the data, `draw(...)`.

 The D3.js cannot read local files (files located on your machine) directly. You will need to set up a web server (viable options of either Apache or Node.js). You need to put the file on the server (in case of Apache) or you can read the data using Node.js and then pass it to D3.js that way. See `https://www.packtpub.com/web-development/instant-nodejs-starter-instant` to start developing with Node.js or search on Google or Bing for numerous online resources to learn how to set up your own (local) web server.

The `draw(...)` function starts with finding the maximum for the price and floor area; this will be used to define the scale and axes of our chart:

```
var limit_max = {
    'price': d3.max(dataset, function(d) {
        return parseInt(d['price']); }),
    'sq__ft': d3.max(dataset, function(d) {
        return parseInt(d['sq__ft']); })
};
```

We use the `.max(...)` method of D3.js. The method returns the maximum element from the passed array. Optionally, one can specify an `accessor` function (as we did here) to access the data in our dataset and tweak it according to one's needs. Our anonymous function, for each element in the `dataset`, returns the element parsed to an integer (as the data read is represented as a string).

Next, we define scales:

```
var scale_x = d3.scale.linear()
    .domain([0, limit_max['price']])
    .range([spacing, width - spacing * 2]);

var scale_y = d3.scale.linear()
    .domain([0, limit_max['sq__ft']])
    .range([height - spacing, spacing]);
```

The `scale.linear()` method of D3.js creates a linearly-spaced scale using passed parameters. The domain specifies the range of our dataset. Our prices fall anywhere between 0 and $884,000 (`limit_max['price']`) and the floor area anywhere between 0 and 5,822 sq. ft (`limit_max['sq__ft']`). The `range(...)` parameter specifies how the domain translates to the size of the SVG window. For `scale_x`, this corresponds to the following relation: if the `price = 0`, then the dot on the chart will be put at 60 pixels from the left, if the price is $884,000, the dot will be placed at 600 (width) - *60 (spacing)* * 2 = 480 pixels from the left.

Now let's define axes:

```
var axis_x = d3.svg.axis()
    .scale(scale_x)
    .orient('bottom')
    .ticks(5);

var axis_y = d3.svg.axis()
    .scale(scale_y)
    .orient('left')
    .ticks(5);
```

The axes need to have a scale and that is what we pass to the axis first. For `axis_x`, we want it to be at the bottom, hence the `.orient('bottom')` parameter; `axis_y` should be tied to the left of the chart (`.orient('left')`). The `.ticks(5)` parameter specifies how many ticks should be present on the axis; D3.js selects the best interval between the ticks automatically based on the data passed.

 You can specify your own tick values by calling `.tickValues(...)` instead of `.ticks(...)`.

We are now ready to plot the dots on the chart:

```
chart.selectAll('circle')
    .data(dataset)
    .enter()
    .append('circle')
    .attr('cx', function(d) {
```

```
        return scale_x(parseInt(d['price']));
    })
    .attr('cy', function(d) {
        return scale_y(parseInt(d['sq__ft']));
    })
    .attr('r', 3)
    ;
```

It starts by selecting all the circles that are on our chart; as there are none, this command returns an empty array. The `.data(dataset).enter()` chain effectively forms a for loop that, for each element of the dataset, appends a dot to our chart, `.append('circle')`. A circle requires three parameters: cx (horizontal position of the circle), cy (vertical position of the circle), and r (its radius). The `.attr(...)` parameters are specifying these parameters for each circle appended to the chart. As you can read from the code, our anonymous functions return translated (scaled) values of the price and floor area.

Having plotted the dots on the chart, it is time to draw the axes:

```
chart.append('g')
    .attr('class', 'axis')
    .attr('transform',
        'translate(0,' + (height - spacing) + ')'
    )
    .call(axis_x);

// Append Y axis to chart
chart.append('g')
    .attr('class', 'axis')
    .attr('transform', 'translate(' + spacing + ',0)')
    .call(axis_y);
```

The documentation of D3.js specifies that, in order to append an axis, we first need to append a g element. We then specify the class for the g element (axis) to make it black and thin (without this, the axis would be thick and so would be ticks—refer to the HTML file for the CSS code that does this). The translation attribute moves the axis from the top of the chart to the bottom, the first parameter specifies the movement along the horizontal axis, and the second parameter specifies the vertical translation. At the end, we call axis_x. Although this might seem odd—why would we call a variable?—the explanation is really simple: axis_x is not really an array of numbers but a function that generates a lot of SVG elements. By calling the function, we append all these elements to the g element that we have just placed on our chart.

Lastly, we want to put the axes' labels on the chart so that people know what the axes represent:

```
chart.append('text')
    .attr("transform",
        "translate(" + (width / 2) + " ," +
```

```
            (height - spacing / 3) + ")"
    )
    .style('text-anchor', 'middle')
    .text('Price $ (,000)');

chart.append("text")
    .attr("transform", "rotate(-90)")
    .attr("y", 14)
    .attr("x",0 - (height / 2))
    .style("text-anchor", "middle")
    .text("Floor area sq. ft.");
```

We append a text field to our chart, specify its position, and anchor the text in the middle. The transform `rotate` turns our label 90 degrees counterclockwise. The `.text(...)` parameter specifies the text of the label.

All these attributes produce our chart:

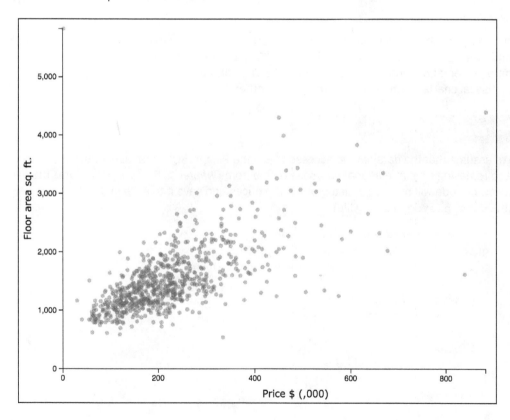

D3.js can be really helpful if you want to visualize your data. A good set of tutorials on D3.js can be found at `http://alignedleft.com/tutorials/d3/`. I also suggest studying Mike Bostock's examples at `https://github.com/mbostock/d3/wiki/Gallery`.

Producing histograms

Histograms can help you inspect the distribution of any variable quickly. Histograms bin the observations into groups and present the count of observations in each group on a bar chart. It is a powerful and easy way to examine issues with your data as well as inspect visually if the data follows some known distribution.

In this recipe, we will produce histograms of all the prices from our dataset.

Getting ready

To run this recipe, you will need `pandas` and `SQLAlchemy` to retrieve the data. `Matplotlib` and `Seaborn` handle the presentation layer. `Matplotlib` is a 2D library for scientific data presentation. Seaborn builds on `Matplotlib` and provides an easier way to produce statistical charts (such as histograms, among others).

How to do it...

We assume that the data can be accessed from the PostgreSQL database. Check the *Storing and retrieving from a relational database* recipe from *Chapter 1, Preparing the Data*. The following code will produce a histogram of the prices and save it to a file in a PDF format (the `data_histograms.py` file):

```
# read prices from the database
query = 'SELECT price FROM real_estate'
price = pd.read_sql_query(query, engine)

# generate the histograms
ax = sns.distplot(
    price,
    bins=10,
    kde=True     # show estimated kernel function
)
```

```
# set the title for the plot
ax.set_title('Price histogram with estimated kernel function')

# and save to a file
plt.savefig('../../Data/Chapter02/Figures/price_histogram.pdf')

# finally, show the plot
plt.show()
```

How it works...

First, we read the data from the database. We omitted the part where we create the engine that connects to our database—refer to either the recipe from the previous chapter or the source code. The price variable is essentially a list of all the prices from our dataset.

Generating histograms with Seaborn is a walk in a park and can be achieved in one line. The `.distplot(...)` method takes a list of numbers (our price variable) as its first (and the only required) parameter. All the other specified parameters are optional. The `bins` parameter specifies how many bins should be created. The `kde` parameter specifies whether the plot should also present estimated kernel density.

 The kernel density estimation is a useful non-parametric technique employed to estimate a **probability density function** (**PDF**) of an unknown distribution. The kernel function has to integrate to 1 (that is, its cumulative density function reaches a maximum of 1 for the whole domain of the function) and has a mean of zero.

The `.distplot(...)` method returns an Axes object (refer to `http://matplotlib.org/api/axes_api.html`) that works as a canvas for our chart. The `.set_title(...)` method creates a title for our chart.

We save our chart using the `.savefig(...)` method of `Matplotlib`. The only parameter required is the location and name of the file to store. The `.savefig(...)` method is smart enough to infer the proper format of the file from the extension specified in the filename. Acceptable filename extensions are `raw` and `rgba` for raw RGBA `bitmap`, `pdf`, `svg`, and `svgz` for Scalable Vector Graphics, `eps` for Encapsulated Postscript, `jpeg` or `jpg`, `bmp`, `png`, `gif`, `pgf` (PGF code for LaTeX), `tif` or `tiff`, and `ps` (Postscript).

The last method sends the chart to the screen:

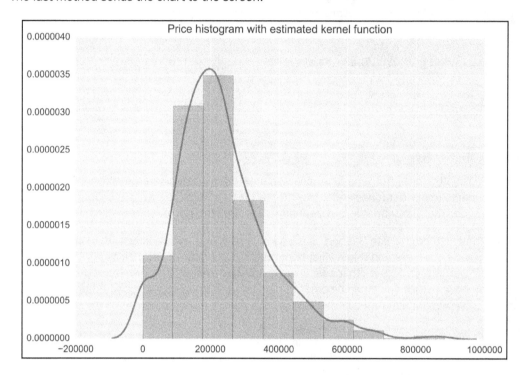

There's more...

Seaborn provides you with a friendly way of creating histograms. However, we can produce (arguably less visually appealing) histograms solely with Matplotlib and pandas. Although Seaborn lets us add many things to our single variable histogram, pandas can produce histograms by groups; here, we present how to produce histograms of prices conditional on the number of bedrooms.

First, we read the data from the CSV file and limit the number of bedrooms to be less than five. (There were not many houses sold with five plus bedrooms so there is little point in producing a histogram for such a group.) This can be achieved using the .query(...) method of pandas. The following code is stored in the data_histograms_alternative.py file:

```
csv_read = pd.read_csv(r_filenameCSV)
csv_read = csv_read.query('beds < 5')
```

Now we can create the histograms using the .hist(...) method.

```
csv_read.hist(
    column='price',
    by='beds',
```

```
            xlabelsize=8,
            ylabelsize=8,
            sharex=True,
            sharey=True,
            figsize=(9,7)
    )
```

First, we specify the `column` (variable) that we want to be presented as a histogram. The `by` parameter specifies the column that we want to produce separate histograms for, in our case, the number of bedrooms. The `xlabelsize` and `ylabelsize` variables specify the size of the x and y label fonts, while `sharex` and `sharey` fix the x and y axes to be equal for all the histograms (so that we can easily compare the distributions). The last parameter, `figsize`, specifies the size of the figure to produce (in inches).

These parameters produce the following chart:

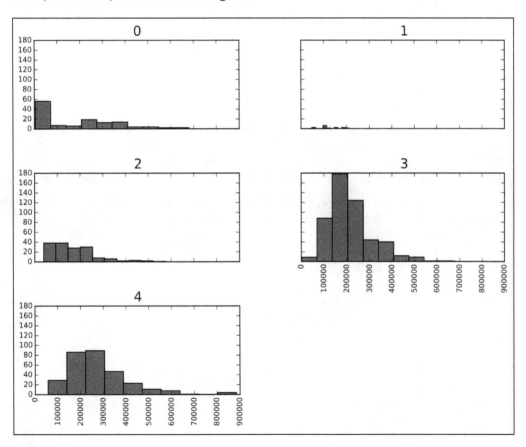

See also...

You can also check another visualization module for Python, `plotly`:

`https://plot.ly/python/histograms/`.

Creating multivariate charts

What the previous recipe showed is that only a handful of houses with less than two bedrooms were sold in the Sacramento area. In the *Visualizing the interactions between features* recipe, we used D3.js to present the relationship between the price and floor area. In this recipe, we will add an another dimension to the two-dimensional chart, the number of bedrooms.

Getting ready

To execute this recipe, you will need the `pandas`, `SQLAlchemy`, and `Bokeh` modules installed. No other prerequisites are required.

How to do it...

`Bokeh` is a module that marries `Seaborn` and D3.js: it produces visually appealing data visualizations (just like `Seaborn`) and allows you to interact with the chart, using D3.js in the backend of the produced HTML file. Producing a similar chart in D3.js would require more coding. The source code for this recipe is contained in the `data_multivariate_charts.py` file:

```
# prepare the query to extract the data from the database
query = 'SELECT beds, sq__ft, price / 1000 AS price \
    FROM real_estate \
    WHERE sq__ft > 0 \
    AND beds BETWEEN 2 AND 4'

# extract the data
data = pd.read_sql_query(query, engine)

# attach the color based on the bed count
data['color'] = data['beds'].map(lambda x: colormap[x])

# create the figure and specify label for axes
fig = b.figure(title='Price vs floor area and bed count')
fig.xaxis.axis_label = 'Price ($ \'000)'
fig.yaxis.axis_label = 'Feet sq'

# and plot the data
```

```
for i in range(2,5):
    d = data[data.beds == i]

    fig.circle(d['price'], d['sq_ft'], color=d['color'],
        fill_alpha=.1, size=8, legend='{0} beds'.format(i))

# specify the output HTML file
b.output_file(
    '../../Data/Chapter02/Figures/price_bed_area.html',
    title='Price vs floor area for different bed count'
)
```

How it works...

First, as always, we need the data; we read the data from the PostgreSQL database in the way you should be familiar with. Then, we attach a color to each record that will help us visualize the relationship between the price and floor area given a different number of bedrooms. The `colormap` variable is specified as follows:

```
colormap = {
    2: 'firebrick',
    3: '#228b22',
    4: 'navy'
}
```

The colors can be specified as readable strings or hexadecimal values (as shown in the preceding code).

To map the number of bedrooms to the specified color, we use lambda. Lambdas are built-in functionality of Python allowing you to specify an unnamed short function to fulfill a single task in-place instead of defining a normal function. It also translates to a more readable code.

 Check this tutorial to understand better why lambdas are helpful and where it is convenient to use them:

https://pythonconquerstheuniverse.wordpress.com/2011/08/29/lambda_tutorial/

Now we can create the figure. We use the `.figure(...)` method of Bokeh. You can specify the title of the chart (as we did) and width and height of the figure. We also define the labels for the axes so that our viewers know what they are looking at.

Then we plot the data. We loop through the list of all possible bedroom numbers `range(2,5)`. (This actually generates a list `[2,3,4]`—refer to the documentation of the `range(...)` method, `https://docs.python.org/3/library/stdtypes.html#typesseq-range`.) For each bedroom count, we extract a subset from the data and store it in the `d` DataFrame.

For each record in the `d` object, we then append a circle to our figure. The `.circle(...)` method takes *x* and *y* coordinates as the first and second parameters. As we want to see the differences in the price-area relationship for different number of bedrooms, we specify the color of the circle. The `fill_alpha` parameter defines a transparency of the circle; it can take values from a range `[0,1]`, where a value of `0` signifies a completely transparent circle and `1` signifies a completely opaque one. The `size` parameter specifies how big the circle should be and `legend` specifies a value to be appended to the chart's legend.

`Bokeh` saves the chart as an interactive HTML file. The `title` parameter specifies the title that is presented when you open the file. It is a fully contained HTML code. The file stores HTML code along with JavaScript and CSS. Our code produces the following chart:

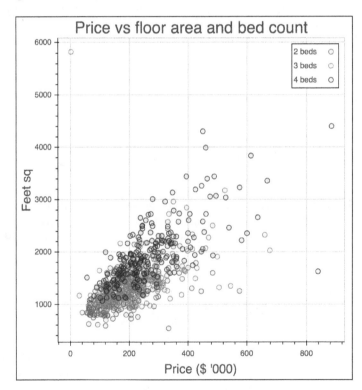

The chart can be panned, zoomed, and resized. Once you are happy with what the chart presents, you can save it to a file:

The icons presented on the screen allow you to (from the left) pan, zoom, and resize the chart window, zoom using your mouse scroll wheel, and save or reset the chart to its original state. The last icon brings up a help window where you can find information on how to add or alter the functionality of these interactive buttons.

See also...

Bokeh is a very powerful visualization library. To glimpse its power, take a look at the examples gallery:

http://bokeh.pydata.org/en/latest/docs/gallery.html

Sampling the data

Sometimes the dataset that we have is too big to be used to build a model. For practical reasons (so that the estimation of our models does not take forever), it is good to create a stratified sample from the full dataset.

In this recipe, we will read from our MongoDB database and use Python to create a sample.

Getting ready

To execute this recipe, you will need PyMongo, pandas, and NumPy. No other prerequisites are required.

How to do it...

There are two approaches that one can take: either specify the fraction of the original dataset (say, 20%) or specify the number of records one would like to retrieve from the dataset. The following code shows you how to fetch a fraction of the dataset (the data_sampling.py file):

```
strata_frac = 0.2

client = pymongo.MongoClient()
db = client['packt']
```

```
real_estate = db['real_estate']

# retrieve the data
sales = pd.DataFrame.from_dict(
    list( real_estate.find(
            {'beds': {'$in': [2,3,4]} },
            {'_id': 0, 'zip': 1, 'city': 1, 'price': 1,
                'beds': 1,'sq__ft': 1}
        )
    )
)

# and select the sample
sample = pd.DataFrame()

for bed in [2,3,4]:
    sample = sample.append(
        sales[sales.beds == bed].sample(frac=strata_frac),
        ignore_index=True
    )
```

How it works...

First, we specify the fraction that we would like to sample from the dataset, the `strata_frac` variable. We retrieve the data from our MongoDB database (see the *Storing in and retrieving from MongoDB* recipe from *Chapter 1, Preparing the Data*, if you cannot understand this part of the code). The MongoDB returns a dictionary. As it is easier to work with DataFrames, we create one using the `.from_dict(...)` method of `pandas`.

The `pandas` DataFrame `.sample(...)` method is a very convenient way to get a subset of the dataset. There is one caveat though: all the observations are equally probable to be selected and we might end up with a sample of a distribution of variables that does not represent that of the full dataset.

To alleviate this, in this simple example, we loop through all the values of bedroom counts and select a sample for such a subset using the `.sample(...)` method. The method allows us to specify the `frac` parameter that returns a sample that only comprises a fraction of the subset (given the bedroom count) of the dataset.

We use the `.append(...)` method of DataFrame: the method takes a DataFrame (in our case, `sample`) and appends another DataFrame to the already existing records. The `ignore_index` parameter, when set to `True`, ignores the index of the DataFrame being appended and maintains the index of the original DataFrame.

There's more...

Sometimes, you might want to specify the number of records to be sampled instead of the fraction of the original dataset. The `.sample(...)` method of `pandas`, described earlier, allows this scenario as well (the `data_sampling_alternative.py` file).

First, we specify the number of records that we would like to sample from the original dataset:

```
strata_cnt = 200
```

In order to preserve the ratios of counts of sales with a different number of bedrooms, we first calculate the expected number of records in each bucket:

```
ttl_cnt = sales['beds'].count()
strata_expected_counts = sales['beds'].value_counts() / \
                    ttl_cnt * strata_cnt
```

The `.count()` method of a DataFrame returns the total count of sales for the whole dataset. We can count the expected number of records for each bedroom count by multiplying the proportion of the 2-, 3-, and 4-bedders in the whole dataset and multiplying this by our `strata_cnt` variable. The `.value_counts()` method returns a record count for each unique value in a specified column (beds, in our case). We then divide each record in this dataset by `ttl_cnt` and multiply by our desired sample size.

To sample, we again use the `.sample(...)` method. However, instead of specifying the `frac` parameter, we specify n, the number of records to retrieve from the subset:

```
for bed in beds:
    sample = sample.append(
        sales[sales.beds == bed] \
        .sample(n=np.round(strata_expected_counts[bed])),
        ignore_index=True
    )
```

Splitting the dataset into training, cross-validation, and testing

To build a statistical model that can be trusted, we need to have confidence that it abstracts the phenomenon that we deal with accurately. To gain such trust, we need to test the model to see if it performs well. To assess the accuracy of our model, we cannot use the same dataset that we used for the training.

In this recipe, you will learn how to split your dataset into two subsets quickly: one that is used solely to train the model and the other one is used to test it.

Getting ready

To execute this recipe, you will need `pandas`, `SQLAlchemy`, and `NumPy`. No other prerequisites are required.

How to do it...

We read our data from the PostgreSQL database and store it in the data DataFrame. Conventionally, we would set aside somewhere between 20-40% of our original dataset for testing purposes. In this example, we select 1/3 of the data (the `data_split.py` file):

```
# specify what proportion of data to hold out for testing
test_size = 0.33

# names of the files to output the samples
w_filenameTrain = '../../Data/Chapter02/realEstate_train.csv'
w_filenameTest  = '../../Data/Chapter02/realEstate_test.csv'

# create a variable to flag the training sample
data['train']  = np.random.rand(len(data)) < (1 - test_size)

# split the data into training and testing
train = data[data.train]
test  = data[~data.train]
```

How it works...

We start by specifying what proportion of the dataset to set aside and where to store our data: the two files will hold our training and testing datasets.

We want to randomly select the records that will end up in the testing dataset. Here, we use `NumPy`'s pseudo-random numbers generator. The `.rand(...)` method generates a list of random numbers with a specified length (`len(data)`). The generated random numbers are between `0` and `1`.

We then compare these numbers with the proportion of the dataset that should be saved for the training (`1 - test_size`): if the generated number is smaller than the training part, then we will keep the record in the training dataset (the `train` column will hold a `True` value); otherwise, it will end up in the file for the testing (a `False` value will be stored in the `train` column).

Splitting of the dataset into training and testing happens in the last two lines of the code. The ~ symbol is the logical NOT operator; therefore, if the value stored in the `train` column is `False`, such negation will return `True`.

There's more...

SciKit-learn offers an alternative way of splitting the dataset. We start by splitting the original dataset into two subsets, one that holds the dependent variable (y) and another that holds all the independent variables (x):

```
# select the independent and dependent variables
x = data[['zip', 'beds', 'sq__ft']]
y = data['price']
```

Now we can perform the split:

```
# and perform the split
x_train, x_test, y_train, y_test = sk.train_test_split(
    x, y, test_size=0.33, random_state=42)
```

The .train_test_split(...) method helps to split the datasets into complementary subsets: the ones that hold the training data and the other one for the testing. In each category, we get two datasets: one that stores the dependent variable and the other one to keep all the independent variables.

3
Classification Techniques

In this chapter, we will cover various techniques that will allow you to classify the outbound call data of a bank. You will learn the following recipes:

- ▶ Testing and comparing the models
- ▶ Classifying with Naïve Bayes
- ▶ Using logistic regression as a universal classifier
- ▶ Utilizing Support Vector Machines as a classification engine
- ▶ Classifying calls with decision trees
- ▶ Predicting subscribers with random tree forests
- ▶ Employing neural networks to classify calls

Introduction

In this chapter, we will be classifying the outbound calls of a bank to see if such a call will result in a credit application. We will use the dataset described in *A Data-Driven Approach to Predict the Success of Bank Telemarketing. Decision Support Systems, Elsevier, 62:22-31, June 2014* by *S. Moro, P. Cortez*, and *P. Rita* found at `http://archive.ics.uci.edu/ml/datasets/Bank+Marketing`.

We transformed the dataset so that we can use it in our models. The data dictionary is located at `/Data/Chapter03/bank_contacts_data_dict.txt`.

Testing and comparing the models

Building statistical models without understanding their effectiveness is a pointless exercise as it gives no indication of whether your model works or not. It also makes it impossible to compare between models in order to choose which one performs better.

In this recipe, we will see how to understand whether your models work well.

Getting ready

To execute this recipe, all you need is `pandas` and **`scikit-learn`**. No other prerequisites are necessary.

How to do it...

`pandas` makes it extremely easy to calculate a suite of test statistics of the performance of your model. We will be using the following code to assess the power of our models (the `helper.py` file at the root of the `Codes` folder):

```python
import sklearn.metrics as mt

def printModelSummary(actual, predicted):
    '''
        Method to print out model summaries
    '''
    print('Overall accuracy of the model is {0:.2f} percent'\
        .format(
            (actual == predicted).sum() / \
            len(actual) * 100))
    print('Confusion matrix: \n',
        mt.confusion_matrix(actual, predicted))
    print('Classification report: \n',
        mt.classification_report(actual, predicted))
    print('ROC: ', mt.roc_auc_score(actual, predicted))
```

How it works...

First, from `scikit-learn`, we import its `metrics` module. Then, we output the overall accuracy of the model. It is calculated by looking at how many times our model agrees with the actual classification `(actual == predicted).sum()` and dividing it by the total size of our testing sample `len(actual) * 100`. This gives us an overall percentage of correctly classified cases. If the distribution of the dependent variable is not even, this metric should not be used to assess the performance of your model.

The `.confusion_matrix` method represents the performance of our model in a very clear way—each row in the matrix represents how many times our model confused (or predicted correctly) the real outcome:

```
Confusion matrix:
[[10154   1816]
 [  559    946]]
```

For example, if we take the first row, we can clearly see that there were *10,154 + 1,816 = 11,970* actual observations that had a class 0. Out of these, *10,154* were classified properly, (we call them true negatives as it indicates that the call did not end up in a credit application) and *1,816* were classified improperly (false positives, meaning the call would have been classified as resulting in a credit application when, in fact, it was not). The second row shows how many times our model predicted the actual outcome: *559* observations were classified incorrectly (false negatives, so the model failed to asses properly that the call resulted in a credit application) and *946* were classified properly (true positives, the number of properly classified calls that resulted in a credit application).

All these counts can be used to calculate several metrics. We use `.classification_report(...)` to generate a full set of these metrics:

```
Classification report:
              precision    recall   f1-score    support
      0.0        0.95       0.85      0.90        11970
      1.0        0.34       0.63      0.44         1505

avg / total      0.88       0.82      0.84        13475
```

Precision is the model's ability to not classify an observation as positive when it is not. It is a ratio of true positives (*946*) to the overall number of positively classified records (*1,816 + 946*): this evaluates to *0.34*, which is fairly low. A corresponding value can also be calculated for the negatives: *10,154 / (10,154 + 559) = 0.95*. The overall precision score is a weighted average of the individual precision scores where the weight is the support. The **support** is the total number of actual observations in each class.

Recall can be viewed as the model's capacity to find all the positive samples. It is a ratio of true positives to the sum of true positives and false negatives: *946 / (946 + 559) = 0.63*. Again, the class 0 can be calculated as the ratio of true negatives (*10,154*) to the sum of true negatives and false positives (*1,816*). The total is a weighted value of precision metrics across all classes, weighted by the support.

F1-score is effectively a weighted amalgam of the precision and recall. It is a ratio of twice the product of precision and recall to their sum. In one measure, it shows whether the model performs well or not.

The last metric we use to assess the model's performance is **Receiver Operating Characteristic (ROC)**. The ROC is a curve that visualizes (for varying levels of true positive rates to false positive rates) performance of the model. In other words, it is effectively a trade-off curve between allowing more false positives to get more true positives. For our purposes, we are more interested in a single metric that is the area under the ROC curve. The models with ROC between 0.9 and 1 are considered excellent, while ROC scores between 0.5 and 0.6 are indicators of a worthless model—a model that is no different from tossing a coin.

There's more...

In our `helper.py` file, we stored a couple of more helpful procedures that we will use heavily in this and the following chapters. The most heavily used will be the `timeit(...)` decorator. We can use the decorator to measure how long they take to execute:

```
def timeit(method):
    '''
        A decorator to time how long it takes to estimate
        the models
    '''

    def timed(*args, **kw):
        start = time.time()
        result = method(*args, **kw)
        end = time.time()

        print('The method {0} took {1:2.2f} sec to run.' \
            .format(method.__name__, end-start))
        return result

    return timed
```

The decorator returns the `timed(...)` method that measures the difference between the timestamp of the end of the execution of a method and its start. To use a decorator, we put `@timeit` (or, as we will see, `@hlp.timeit`) just before the method that we want to measure the execution time of. The `timeit(...)` method's only argument is `method` that gets passed on to the `timed(...)` internal method. The `timed(...)` method starts the timer, executes the method passed, and prints out how long it took.

Functions in Python are nothing more than objects that can be passed around just like another object (for example, `int`). Therefore, we can pass a method to another function as a parameter and then use it inside the other method. More on decorators (for Python 2.7) can be found at `http://thecodeship.com/patterns/guide-to-python-function-decorators/`.

See also

I strongly suggest studying `Scikit`'s documentation with regard to various metrics for classification models:

`http://scikit-learn.org/stable/modules/model_evaluation.html`

Classifying with Naïve Bayes

The Naïve Bayes classifier is one of the simplest classification techniques. It uses Bayes' Theorem of conditional probability of an event happening given that some other event occurs. The Naïve Bayes classifier leverages the very familiar formula:

$P(A|B) = P(B|A)\, P(A)\, /\, P(B)$

In other words, we want to calculate the probability of the outbound call resulting in a credit application (*A*) given the various characteristics of the call and caller (*B*). This is equivalent to the ratio of the product of the observed frequency of applying for credit: the *P(A)*, and the frequency of such characteristics of the call and caller occurring for those who had taken the offer in the past: the *P(B|A)*, to the frequency of such a call and caller occurring in our dataset: the *P(B)*.

Getting ready

To execute this recipe, you will need `pandas` and `scikit-learn`. We will also use our `helper.py` script so you will need `NumPy` and `time` modules.

As our `helper.py` is located in the parent folder, we need to add the parent folder to Python's path so that Python will search for codes there:

```
# this is needed to load helper from the parent folder
import sys
sys.path.append('..')

# now you can import the script
import helper as hlp
import pandas as pd
import sklearn.naive_bayes as nb
```

How to do it...

Building a Naïve Bayes classifier in `pandas` takes two lines of code. Preparing for that stage takes a bit longer (the `classification_naiveBayes.py` file):

```
# the file name of the dataset
r_filename = '../../Data/Chapter03/bank_contacts.csv'

# read the data
csv_read = pd.read_csv(r_filename)

# split the data into training and testing
train_x, train_y, \
test_x,  test_y, \
labels = hlp.split_data(
    csv_read, y = 'credit_application')

# train the model
classifier = fitNaiveBayes((train_x, train_y))

@hlp.timeit
def fitNaiveBayes(data):
    '''
        Build the Naive Bayes classifier
    '''
    # create the classifier object
    naiveBayes_classifier = nb.GaussianNB()

    # fit the model
    return naiveBayes_classifier.fit(data[0], data[1])
```

How it works...

We start with reading in the data from our CSV file in a very familiar way (refer to the *Reading and writing CSV/TSV files with Python* recipe from *Chapter 1, Preparing the Data*). Then, we split the data into training and testing, keeping our independent (x) and dependent (y) variables separate (refer to *Splitting the dataset into training and testing* recipe from *Chapter 2, Exploring the Data*).

The `.split_data(...)` method from our `helper.py` script splits the dataset into two subsets: by default, it uses 2/3 of the dataset to train and 1/3 to test. The method is a slightly more complex cousin of the method that we developed in the previous chapter. The two required parameters are (in order) `data` and `y` (our dependent variable). The `data` parameter should be of the `pandas` DataFrame type, whereas `y` should be a column name from this DataFrame. You can also select the independent variables that you want to keep (by default, all the columns except the one specified as `y` are used) by passing a list of column names to the `x` parameter. (We will use this in the recipes later in this chapter.) You can also alter the proportion that the method should use for the testing by specifying the `test_size` parameter and passing a number between `0` and `1`.

Now that we have the dataset split into testing and training, we can build the classifier. In this and the following recipes, we will be timing how long it takes to build different models using our `@hlp.timeit` decorator.

To build our Naïve Bayes classifier, we use the `.GaussianNB()` classifier from `Scikit`. The method takes no additional arguments. To fit the model, we use the `.fit(...)` method; the first argument passed is the set of independent variables, the second one is the vector with the dependent variable. Once fit, we return the classifier to the main script. We are done with building the model!

Now we need to test the model. The `.predict(...)` method uses the `test_x` dataset and classifies the observations into two buckets: either the call will result in a credit application or not.

 `Scikit` models follow the same naming patterns across models; that is why the framework is so convenient to use. You will see that other models that we will use later all have the `.fit(...)` and `.predict(...)` methods, making it extremely easy to test various models without having to rewrite extended portions of your code.

To test the model's performance, we use the previously introduced `printModelSummary(...)` method (refer to the *Testing and comparing the models* recipe). The `fitNaiveBayes(...)` method, on an average, took 0.03 seconds to finish (on my machine).

 We quote the time here that most likely will differ when you run the script on your computer. We (and so shall you) will only be using this as a benchmark to compare the methods that we will introduce later.

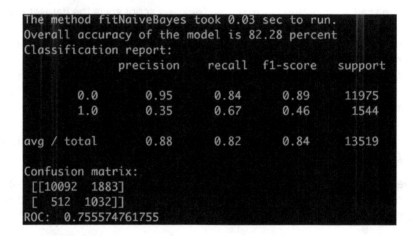

```
The method fitNaiveBayes took 0.03 sec to run.
Overall accuracy of the model is 82.28 percent
Classification report:
                precision    recall  f1-score   support

        0.0         0.95      0.84      0.89     11975
        1.0         0.35      0.67      0.46      1544

avg / total         0.88      0.82      0.84     13519

Confusion matrix:
 [[10092  1883]
 [  512  1032]]
ROC:   0.755574761755
```

See also

Naïve Bayes is a widely used machine learning technique with multitude of applications. See, for example, how one can use it to classify text at `http://sebastianraschka.com/Articles/2014_naive_bayes_1.html`.

Using logistic regression as a universal classifier

Logistic regression is probably the second most (after linear regression) popular regression model. However, it can be easily adapted to solve a classification problem.

Getting ready

To run this recipe, you will need `pandas` and `StatsModels`; if you use the Anaconda distribution of Python, both of the modules are included in the distribution. We import two parts of `StatsModels`:

```
import statsmodels.api as sm
import statsmodels.genmod.families.links as fm
```

The first one allows us to select our models and the other one to specify the link function. No other prerequisites are required.

How to do it...

Following a similar pattern to our previous recipe, we import all the necessary modules first, read in the data, and split the read dataset into training and testing subsets. We then call the `fitLogisticRegression(...)` method to estimate the model (the `classification_logistic.py` file):

```
@hlp.timeit
def fitLogisticRegression(data):
    '''
        Build the logistic regression classifier
    '''
    # create the classifier object
    logistic_classifier = sm.GLM(data[1], data[0],
        family=sm.families.Binomial(link=fm.logit))

    # fit the data
    return logistic_classifier.fit()
```

How it works...

The `StatsModels` module provides the `.GLM(...)` method. **GLM** stands for a **Generalized Linear Model**. It is a family of models that generalize a linear regression (that assumes the normality of the error terms) to other distributions. The model can be expressed as follows:

$E(Y) = g^{-1}(X\beta)$

Here, g is a link function, $X\beta$ is the set of linear equations, and $E(Y)$ is the expectation of the dependent variable (think of it as an average of a non-normal distribution). The GLM uses the link function to relate the model to the distribution of the response variable.

 For a list of all link functions, refer to `http://statsmodels.sourceforge.net/devel/glm.html#links`.

The `Statsmodels'` `.GLM(...)` method allows us to specify a multitude of different distributions. In our case, we are fitting a binomial variable, a variable with only two possible outputs: 1 signifying a call ending up in a credit application and 0 means otherwise. Hence, we use `.families.Binomial(...)`. The binomial family's default link function is, in fact, the `logit` function so we did not have to specify the link explicitly; we defined it so you know how to do it should you decide to use other link functions in your models.

See `http://statsmodels.sourceforge.net/` `devel/glm.html#families` for a list of all model families implemented in StatsModels.

The `logit` function is an inverse of the more familiar `sigmoid` function (see the following image):

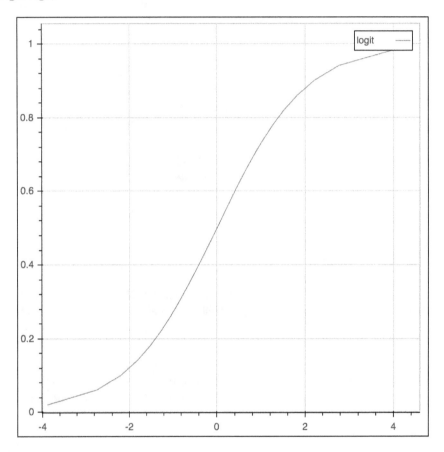

As you can see, the output from the logistic regression can only vary between 0 and 1 (inclusive) and we can treat it as a probability of the call transforming into a credit application; once it is greater than 50%, we treat it as more likely and assign it to the class 1, otherwise, 0. This can be attained with the following code:

```
# classify the unseen data
predicted = classifier.predict(test_x)

# assign the class
```

```
predicted = [1 if elem > 0.5 else 0 for elem in predicted]

# print out the results
hlp.printModelSummary(test_y, predicted)
```

We use list comprehension to assign the final class to our testing dataset. Then, we use our `.printModelSummary(...)` method to outline the model's performance. The performance of the logistic regression classifier is better than Naïve Bayes', but this comes at a significantly higher computation cost.

```
The method fitLogisticRegression took 2.02 sec to run.
Overall accuracy of the model is 91.08 percent
Classification report:
                precision     recall   f1-score     support

          0.0        0.93       0.97       0.95       12106
          1.0        0.68       0.42       0.52        1565

avg / total          0.90       0.91       0.90       13671


Confusion matrix:
 [[11788   318]
 [  901   664]]
ROC:   0.699006591931
```

Compared to the Naïve Bayes classifier, we got better at classifying true positives at the cost of overstating false positives.

The GLM classifier of `StatsModels` allows us to print out more detailed results of the regression itself, along with coefficient values and their statistical performance. The `.summary()` method invoked on our classifier will print something similar (abbreviated) to the following image:

```
                  Generalized Linear Model Regression Results
================================================================================
Dep. Variable:     credit_application   No. Observations:              27517
Model:                            GLM   Df Residuals:                  27465
Model Family:                Binomial   Df Model:                         51
Link Function:                  logit   Scale:                           1.0
Method:                          IRLS   Log-Likelihood:              -5776.2
Date:                Mon, 14 Mar 2016   Deviance:                     11552.
Time:                        21:09:05   Pearson chi2:                5.25e+07
No. Iterations:                    22
================================================================================
                coef      std err          z      P>|z|      [95.0% Conf. Int.]
--------------------------------------------------------------------------------
n_age         0.1918        0.240      0.798      0.425       -0.279      0.663
n_duration   22.5235        0.441     51.024      0.000       21.658     23.389
n_pdays      -1.0472        0.273     -3.832      0.000       -1.583     -0.512
n_previous   -0.2053        0.505     -0.406      0.685       -1.196      0.785
```

You can clearly see that the n_age variable is insignificant and can be dropped from the model without losing much accuracy, while n_duration is both, highly significant and has a huge influence on whether the customer applies for credit or not. Such information helps shape your model and make it better defined by removing irrelevant (and potentially error-introducing) variables.

There's more...

You can also use Scikit's method to estimate a logistic regression classifier (the classification_logistic_alternative.py file):

```
import sklearn.linear_model as lm

@hlp.timeit
def fitLogisticRegression(data):
    '''
        Build the logistic regression classifier
    '''
    # create the classifier object
    logistic_classifier = lm.LogisticRegression()

    # fit the data
    return logistic_classifier.fit(data[0], data[1])
```

Although the method is quicker than that of StatsModels, it comes at a cost of not being able to assess which of the coefficients are statistically significant and which are not; the .LogisticRegression(...) classifier does not produce such statistics.

The model produces similar results in terms of accuracy as the one of StatsModels, trading the precision over recall when compared with the Naïve Bayes classifier.

See also

In the next chapter, we will introduce mlpy, a machine learning module. It can also be used to estimate linear models:

http://mlpy.sourceforge.net/docs/3.5/liblinear.html

Utilizing Support Vector Machines as a classification engine

Support Vector Machines (**SVMs**) are a family of extremely powerful models that can be used in classification and regression problems. In contrast to the preceding models, SVMs can handle highly nonlinear problems through a so-called kernel trick that implicitly maps the input vectors to higher-dimensional feature spaces. A broader explanation of SVMs can be found at http://www.statsoft.com/Textbook/Support-Vector-Machines.

Getting ready

To execute the following recipe, you will need **Machine Learning PYthon** (**mlpy**). The mlpy does not come with Anaconda so we need to install it manually. The mlpy requires **GNU Scientific Library** (**GSL**); on some systems, GSL might already be present, therefore, I recommend starting with installing mlpy first. Go to http://sourceforge.net/projects/mlpy/files/ and download the latest sources for mlpy (mlpy-<version>.tar.gz). Now, go to the command line and navigate to the folder you have downloaded the source file to (the cd command is your ally).

 On Mac OS X, if you download the files using Safari, it will automatically unarchive .gz and you will see .tar files.

If you see files with a .gz extension, issue the following command:

```
gzip -c mlpy-<version>.tar.gz | tar -xv
```

If, however, the file has already been extracted and you see a .tar extension, use the following:

```
tar -xvf mlpy-<version>.tar
```

Either of these commands will create a mlpy-<version> folder.

 The | symbol helps execute multiple commands in sequence. The syntax is called piping. You can read more about it at http://www.westwind.com/reference/os-x/commandline/pipes.html.

The preceding commands should be read from left to right. Each command is separated by the | (pipe). The execution starts with the first command from the left and the results of that operation are then used as an input to the next command in the chain.

Now, navigate to that folder and issue the following command:

```
python setup.py install
```

In case the preceding command fails with the following error:

```
mlpy/gsl/gsl.c:223:10: fatal error: 'gsl/gsl_sf.h' file not found
```

Then, you will need to install GSL.

To do this, first go to http://gnu.mirror.iweb.com/gsl/ and download the latest version of GSL (scroll to the bottom and download gsl-latest.tar.gz).

To extract the contents of the archives, issue the following command (see the previous note if the file is already in a .tar format):

```
gzip -c gsl-latest.tar-gz | tar -xv
```

Navigate to the gsl-latest folder and issue the following command:

```
./configure
```

This will scan your computer and configure the compilation process for you. Then type the following:

```
make
```

```
make install
```

The preceding commands will compile the source codes for GSL and install it on your computer. Assuming that all goes well, go back to the mlpy folder and install the mlpy; it should now get installed without any problems.

However, sometimes GSL is not installed in the default include directory. On Mac OS X, you can try and look for the gsl folder in /usr/local/include/ (for header files) and /usr/local/lib/ (for the compiled library). Once you find the folder that GSL is installed in, then issue the following command to install mlpy (assuming that the headers and libraries are located in the folders mentioned). The following command should be typed in one line:

```
python setup.py build_ext --include-dirs=/usr/local/include/ --rpath=/usr/local/lib
```

How to do it...

Building an SVM classifier with the mlpy is easy. All you need to do is specify the SVM type and its kernel (the classification_svm.py file):

```
import mlpy as ml

@hlp.timeit
def fitLinearSVM(data):
    '''
```

```
    Build the linear SVM classifier
'''
# create the classifier object
svm = ml.LibSvm(svm_type='c_svc',
    kernel_type='linear', C=100.0)

# fit the data
svm.learn(data[0],data[1])

# return the classifier
return svm
```

How it works...

First, we load the `mlpy`. Building the model starts with calling the `.LibSvm(...)` method. The first parameter that we specify is `svm_type`. The `c_svc` specifies a C-support Vector Classifier; the C parameter penalizes the error term. You can also specify `nu_svc` with a nu parameter that controls how much of your sample (at most) can be misclassified and how many of your observations (at least) can become support vectors.

The `kernel_type` specifies, as the name suggests, the type of the kernel. We can choose either linear, polynomial, **Radial Basis Functions** (**RBFs**), or sigmoid. Depending on our choice of the kernel, the model will try to fit a line, polyline, or nonlinear shapes to find the best separation of our classes.

 While linear kernels are well-suited for linearly separable problems, RBFs can be used to find highly-nonlinear boundaries between classes. Check this presentation at `http://www.cs.cornell.edu/courses/cs578/2003fa/slides_sigir03_tutorial-modified.v3.pdf`.

Once we define the model, we allow it to `.learn(...)` the support vectors from our data. To fit an RBF version of our model, we can specify our `svm` object as follows:

```
svm = ml.LibSvm(svm_type='c_svc', kernel_type='rbf',
    gamma=0.1, C=1.0)
```

The `gamma` parameter here specifies how far the influence of a single support vector reaches. Visually, you can investigate the relationship between gamma and C parameters at `http://scikit-learn.org/stable/auto_examples/svm/plot_rbf_parameters.html`.

SVMs, of all the classification methods presented so far, are the slowest to estimate. Also, they do not perform as well as the logistic regression classifier. This, however, should not be treated as any indication that logistic regression is a panacea for all your classification problems. What I am trying to note here is that, for this particular dataset, the logistic regression classifier performed better than SVM; for some more nonlinear problems, SVMs may be more suitable and will outperform the logistic regression classifier. This can also be observed from our SVM tests: the SVM with the linear kernel performed better than its RBF counterpart:

```
The method fitLinearSVM took 100.20 sec to run.
The method fitRBFSVM took 14.02 sec to run.
Overall accuracy of the model is 90.58 percent
Classification report:
             precision    recall  f1-score   support

        0.0       0.92      0.98      0.95     12113
        1.0       0.65      0.32      0.43      1514

avg / total       0.89      0.91      0.89     13627


Confusion matrix:
 [[11853   260]
 [ 1023   491]]
ROC:  0.651420965346
Overall accuracy of the model is 89.70 percent
Classification report:
             precision    recall  f1-score   support

        0.0       0.91      0.99      0.94     12113
        1.0       0.63      0.18      0.28      1514

avg / total       0.88      0.90      0.87     13627


Confusion matrix:
 [[11955   158]
 [ 1245   269]]
ROC:  0.582315597913
```

There's more...

Scikit-learn also provides you with a method to estimate SVMs (the classification_svm_alternative.py file):

```
import sklearn.svm as sv

@hlp.timeit
def fitSVM(data):
    '''
        Build the SVM classifier
```

```
'''
# create the classifier object
svm = sv.SVC(kernel='linear', C=20.0)

# fit the data
return svm.fit(data[0],data[1])
```

We use a linear kernel in this example but, of course, you can use RBF (which is, in fact, the default setting of `.SVC(...)`). The method produces similar results but is faster than the method provided by `mlpy`. You can also list all of the support vector information if you would like; these are stored in the `.support_vectors_` attribute of classifier:

```
The method fitSVM took 71.98 sec to run.
Overall accuracy of the model is 90.39 percent
Classification report:
              precision   recall  f1-score   support

         0.0       0.92     0.98      0.95     12056
         1.0       0.67     0.31      0.42      1549

avg / total         0.89     0.90      0.89     13605

Confusion matrix:
 [[11816    240]
 [ 1068    481]]
ROC:   0.645307908906
```

Classifying calls with decision trees

Decision trees have been widely used to solve classification problems. A decision tree is, as the name suggests, a tree-like structure that branches out from the root. At each branch (decision) node, the remaining data is split into two groups given a decision criterion with a specified objective. The process continues until no more divisions can be made or all the samples in the ending node (a leaf) belong to the same class (that is, variance is minimized).

Getting ready

To execute this recipe, you will need `pandas` and `Scikit-learn`. To execute an alternate way of estimating a decision tree classifier, you will need `mlpy`. No other prerequisites are required.

How to do it...

Scikit-learn provides the `DecisionTreeClassifier(...)` class that we will use to estimate our decision tree classifier (the `classification_decisionTree.py` file):

```
import sklearn.tree as sk

@hlp.timeit
def fitDecisionTree(data):
    '''
        Build a decision tree classifier
    '''
    # create the classifier object
    tree = sk.DecisionTreeClassifier(min_samples_split=1000)

    # fit the data
    return tree.fit(data[0],data[1])
```

How it works...

First, we import the `sklearn.tree` module that exposes the `DecisionTreeClassifier(...)` class for us to use. Then, we read in the data from the CSV file and split the dataset into training and testing. We have decided that we will use only a subset of all the variables. This can be done by passing a list with all the variables that we want to use as an x parameter to our `.split_data(...)` method:

```
# split the data into training and testing
train_x, train_y, \
test_x,  test_y, \
labels = hlp.split_data(
    csv_read,
    y = 'credit_application',
    x = ['n_duration','n_nr_employed',
        'prev_ctc_outcome_success','n_euribor3m',
        'n_cons_conf_idx','n_age','month_oct',
        'n_cons_price_idx','edu_university_degree','n_pdays',
        'dow_mon','job_student','job_technician',
        'job_housemaid','edu_basic_6y']
)
```

Note that our `.split_data(...)` method returns not only the training and testing subsets of our read dataset, but also returns labels for each of the variables.

Now that we have the training and testing datasets, we can fit the decision tree:

```
classifier = fitDecisionTree((train_x, train_y))
```

The `DecisionTreeClassifier(...)` class can be tweaked in multiple ways. Here, we only specify that any decision node cannot hold less than 1,000 observations.

 For a full list of `DecisionTreeClassifier(...)` parameters, consult

`http://scikit-learn.org/stable/modules/generated/sklearn.tree.DecisionTreeClassifier.html`

Having the `model .fit(...)` to our data, we can then classify the unseen observations from the `test_x` dataset and print out the results:

```
The method fitDecisionTree took 0.06 sec to run.
Overall accuracy of the model is 90.89 percent
Classification report:
                precision    recall  f1-score   support

        0.0         0.94      0.96      0.95     12050
        1.0         0.62      0.54      0.57      1554

avg / total         0.90      0.91      0.91     13604

Confusion matrix:
 [[11526    524]
 [  716    838]]
ROC:   0.747884031038
```

So far, this is the best model that we were able to fit, with good precision and recall. Even though the Naïve Bayes had a similar ROC score, its precision was much worse (see the *Classifying with Naïve Bayes* recipe). Moreover, estimating the decision tree classifier is only slightly slower than the Naïve Bayes classifier.

`Scikit` also provides a useful `.export_graphviz(...)` method that allows you to save the model in a `.dot` format; the `.dot` format is a native format for GraphViz:

```
sk.export_graphviz(classifier,
    out_file='../../Data/Chapter03/decisionTree/tree.dot')
```

The `.dot` format is essentially a text file, not really intuitive to read. However, we can visualize it. To do so, you can either use GraphViz itself or, if you are running Linux or Mac OS X, you have the `dot` tool at your disposal.

 GraphViz is an open source graph visualization tool. You can download it from `http://www.graphviz.org/Download.php`.

To produce a PDF file with our tree, you can execute the following command:

```
dot -Tpdf tree.dot -o tree.pdf
```

The `dot` command can output the tree in a variety of formats: PNG (by specifying `-Tpng`) or SVG (Scalable Vector Graphics, `-Tsvg`) among others.

 For a full list of the formats, refer to `http://www.graphviz.org/content/output-formats`.

The `-o` parameter specifies the output filename (in our case, `tree.pdf`). The output looks as follows (abbreviated):

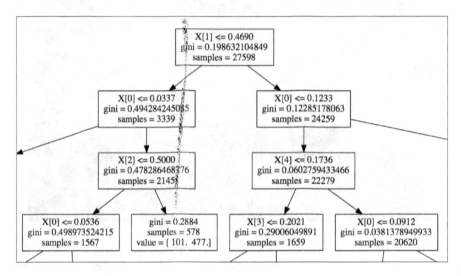

In the graph, each node carries important information about the decision tree:

X[1] <= 0.4690 gini = 0.201481912938 samples = 27600	gini = 0.0853 samples = 829 value = [792. 37.]

The decision node is presented on the left while the final leaf is shown on the right. The **X[1]** specifies the variable from our sample. To track which variable we are splitting first, we can print out the variable names, their position in our dataset, and their importance:

```
for counter, (nm, label) \
    in enumerate(
        zip(labels, classifier.feature_importances_)
    ):
    print("{0}. {1}: {2}".format(counter, nm, label))
```

The preceding code uses the `enumerate(...)` method that returns two elements: the counter and a tuple `(nm, label)`. The `zip(...)` method takes our labels object and the `.feature_importances_` attribute of our decision tree classifier and creates an entity that matches the elements from the labels object with `.feature_importances_` based on their position, that is, the first element of labels with the first element of `.feature_importances_`. Our script generates the following output (abbreviated):

```
0. n_duration: 0.5081646778462993
1. n_nr_employed: 0.35055350868467067
2. prev_ctc_outcome_success: 0.029489215923603578
3. n_euribor3m: 0.035240121468937555
4. n_cons_conf_idx: 0.03581315133871834
5. n_age: 0.015445054892527188
6. month_oct: 0.017559494426098093
```

Now, you can see that our **X[1]** variable is in fact `n_nr_employed`. So, if `n_nr_employed` is less than or equal to `0.4690`, we will follow the tree to the left to the next decision node; otherwise, we would go right.

The `gini` attribute of the decision node indicates the Gini impurity measure. It measures how often a randomly chosen element from the dataset would be classified incorrectly: the closer to `0`, the more confident you are that you are not misclassifying your observations.

> To learn how to calculate the Gini impurity measure, refer to
> http://people.revoledu.com/kardi/tutorial/
> DecisionTree/how-to-measure-impurity.htm.

In the decision node, the `samples` attribute indicates how many samples are being split. The value indicates how many samples fall into each class in the leaf node.

There's more...

The `mlpy` framework also allows us to estimate a decision tree classifier (the `classification_decisionTree_alternative.py` file):

```
import mlpy as ml

@hlp.timeit
def fitDecisionTree(data):
    '''
        Build a decision tree classifier
    '''
    # create the classifier object
    tree = ml.ClassTree(minsize=1000)
```

```
# fit the data
tree.learn(data[0],data[1])

# return the classifier
return tree
```

As with `Scikit`'s `DecisionTreeClassifier`, we specify the `minsize` parameter of the `ClassTree(...)` class. The `ClassTree(...)` classifier has less options to tweak when compared to `Scikit` but produces an equally effective decision tree classifier:

```
The method fitDecisionTree took 0.77 sec to run.
Overall accuracy of the model is 91.50 percent
Classification report:
                  precision    recall  f1-score   support

          0.0        0.95      0.96      0.95     12326
          1.0        0.64      0.56      0.60      1551

avg / total          0.91      0.92      0.91     13877

Confusion matrix:
 [[11827    499]
 [  680    871]]
ROC:   0.760544823923
```

Predicting subscribers with random tree forests

Random forests belong to a family of ensemble models. The ensemble models work on a premise that two brains are better than one; they combine the predictions of many weaker models (decision trees) to come up with a prediction that reflects a mode among these weaker models. For more, check https://www.stat.berkeley.edu/~breiman/RandomForests/cc_home.htm.

Getting ready

To execute this recipe, you will need `pandas` and `scikit-learn`. No other prerequisites are required.

How to do it...

As in previous examples, `Scikit` provides an easy way of building a random forest classifier
(the `classification_randomForest.py` file):

```
import sklearn.ensemble as en

@hlp.timeit
def fitRandomForest(data):
    '''
        Build a random forest classifier
    '''
    # create the classifier object
    forest = en.RandomForestClassifier(n_jobs=-1,
        min_samples_split=100, n_estimators=10,
        class_weight="auto")

    # fit the data
    return forest.fit(data[0],data[1])
```

How it works...

First, we import the necessary module of `scikit-learn` that exposes the
`RandomForestClassifier(...)` class. On reading the dataset and splitting it into training
and testing samples, we estimate `RandomForestClassifier(...)`.

> `RandomForestClassifier(...)` can take a multitude
> of parameters; consult `http://scikit-learn.org/`
> `stable/modules/generated/sklearn.ensemble.`
> `RandomForestClassifier.html` for more information.

First, we specify `n_jobs`. This specifies how many jobs Python should run in parallel during
estimation and prediction. It can make a significant difference if you have a big dataset with
many observations and features and estimate thousands of trees in the ensemble. In our
example, though, switching between `-1` (it spins up as many jobs as the number of cores of
your processor) and `1` (a single job) did not bring any substantial differences.

The `min_samples_split` parameter, just like the decision tree classifier, controls the minimum
number of observations in the decision node to perform a split. The `n_estimators` specifies
how many weak models to build; in our case, we build 10 but these can go into thousands if
required. The `class_weight` parameter controls the weight of each class. This is useful if the
frequencies of the classes in your input data are heavily skewed (as in our case). The `auto` value
for this parameter sets `class_weight` to the inverse of the class frequencies. You can specify
`class_weight` yourself: the parameters accept a dictionary of a `{class: weight}` form.

The `.fit(...)` method can also accept a `sample_weight` parameter (not stated in our code). This specifies the weight of each observation; thus, it needs to be a vector (list) with the length equal to the number of rows in your dataset. It can be useful if you had many observations that you could not trust for whatever reason; instead of completely discarding such observations, you can still include them in your model but with a smaller weight.

The time to estimate `RandomForestClassifier(...)` depends on many factors: the size of your dataset, number of classifiers, and number of jobs you specified. In our case, due to the way it was configured, it did not take too long when compared with decision tree classifiers. By far, this model scored the best in terms of recall and ROC. The precision is not as good as with the other models; the model produces more false positives than true ones:

```
The method fitRandomForest took 0.12 sec to run.
Overall accuracy of the model is 85.55 percent
Classification report:
                 precision     recall   f1-score    support

         0.0        0.99       0.85       0.91       12054
         1.0        0.44       0.93       0.59        1541

avg / total          0.93       0.86       0.88       13595

Confusion matrix:
 [[10203  1851]
 [  114  1427]]
ROC:   0.886231539513
```

We output all the trees to the `/Data/Chapter03/randomForest` folder. In the folder, you can find the `convertToPdf.sh` bash script that automates the conversion from `.dot` to `.pdf`. The script should work fine in any Unix-like environment (Linux, Mac OS X, Cygwin, and others). To run it, issue the following command in the folder:

`./convertToPdf.sh`

The script is a very simple one:

```
#/bin/bash
for f in *.dot;
do
    echo Processing $f;
    dot -Tpdf $f -o ${f%.*}.pdf;
done
```

The first line specifies where the script should be looking for bash. The path might be different on your machine. To find where (and if) you have bash installed, issue the following command:

```
sudo find / -name bash
```

The command will ask you to type in the administrator's password. This will list all the locations where the bash file is located. Look for a path where bash sits in a `bin` folder.

Next, the for loop goes through all the files that have a `.dot` extension; each file's name is stored in the `f` variable. In bash scripting, we use `$f` to access the value stored in the `f` variable. The `echo` command prints out the name of the file we are currently processing to the screen. Then, we use the already familiar `dot` command to convert our `.dot` file into a PDF document. Note how we extract only the name of our file, discarding the `.dot` extension and adding the `.pdf` as the new format.

There's more...

Even though this is not a random forest classifier, the gradient boosted classifier belongs to the same family of models. The way a gradient boosted tree works is very similar to the random forest—both approaches combine weaker models to predict the class. The differences are that random forest models train the trees randomly while gradient boosting attempts at optimizing the linear combination of the trees.

Scikit provides the `GradientBoostingClassifier(...)` class to fit gradient boosted trees (the `classifier_gradientBoosting.py` file):

```
@hlp.timeit
def fitGradientBoosting(data):
    '''
        Build a gradient boosting classier
    '''
    # create the classifier object
    gradBoost = en.GradientBoostingClassifier(
        min_samples_split=100, n_estimators=10)

    # fit the data
    return gradBoost.fit(data[0],data[1])
```

The set of parameters for `GradientBoostingClassifier(...)` is similar to that of `RandomForestClassifier(...)`; we also specify the minimal number of samples for a split, and we set the number of weak models we set to 10.

The gradient boosting classifier, for our data, performed worse than the random forest in terms of recall and ROC, but it outperformed it in terms of precision:

```
The method fitRandomForest took 0.12 sec to run.
Overall accuracy of the model is 85.55 percent
Classification report:
                precision      recall   f1-score     support

          0.0        0.99        0.85       0.91       12054
          1.0        0.44        0.93       0.59        1541

avg / total          0.93        0.86       0.88       13595

Confusion matrix:
 [[10203  1851]
  [  114  1427]]
ROC:   0.886231539513
```

Employing neural networks to classify calls

Artificial Neural Networks (**ANN**) are machine learning models that try to mimic the functions of a biological brain. A fundamental unit of a neural network is a structure called neuron. A neuron has one or multiple inputs and a soma: a part of the neuron that sums up the input signals that are then passed through an activation (transition) function that decides whether (and/or how) to propagate the signal to the output. There is a multitude of various transition functions that an artificial neuron can implement. These range from the basic ones such as the step function that sends a signal only if a certain threshold is exceeded through linear activation functions that do not alter the signal in any way to some nonlinear functions such as tanh, sigmoid, or RBF.

A neural network is a structure that combines such neurons into layers. The input layer is an interface between the neural network and training dataset. Necessarily, it needs to have as many input neurons as the number of x variables. The hidden layer(s) consist of multiple neurons. The inputs of these neurons can be either connected to one, some, or all the preceding layer neurons; these connections have a weight that is applied to the signal before it reaches the next layer, either amplifying or damping it. The number of neurons in the output layer should equal the number of levels that your output variable has. In our case, we will have two neurons as our dependent variable has two levels: whether someone applies for a credit or not. A typical layout of a network with two hidden layers is presented here:

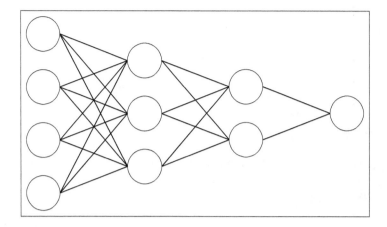

Training of a neural network is a matter of changing the connecting weights between the neurons and altering the activation function parameters of each neuron. The most popular supervised learning paradigm is the error back-propagation training method. The algorithm calculates the error between the network's output and target variable, and then back-propagates this error to the previous layers where the connections and neuron parameters are being adapted depending on how much influence that particular neuron has on the output.

Getting ready

To execute this recipe, you will need pandas and PyBrain. To install PyBrain, execute the following:

cd ~/Downloads

git clone git://github.com/pybrain/pybrain.git

cd pybrain

python setup.py install

How to do it...

Estimating a simple neural network with PyBrain is fairly simple:

```
import pybrain.structure as st
import pybrain.supervised.trainers as tr
import pybrain.tools.shortcuts as pb

@hlp.timeit
def fitANN(data):
    '''
        Build a neural network classifier
```

```
'''
# determine the number of inputs and outputs
inputs_cnt = data['input'].shape[1]
target_cnt = data['target'].shape[1]

# create the classifier object
ann = pb.buildNetwork(inputs_cnt,
    inputs_cnt / 2,
    target_cnt,
    hiddenclass=st.TanhLayer,
    outclass=st.SoftmaxLayer,
    bias=True
)

# create the trainer object
trainer = tr.BackpropTrainer(ann, data,
    verbose=True, batchlearning=False)

# and train the network
trainer.trainUntilConvergence(maxEpochs=50, verbose=True,
    continueEpochs=3, validationProportion=0.25)

# and return the classifier
return ann
```

How it works...

First, we load all the necessary modules from `PyBrain`. The `.structure` gives us access to a variety of activation functions (see `http://pybrain.org/docs/api/structure/modules.html` for more). The `.supervised.trainers` module exposes the supervised methods to train our network (`http://pybrain.org/docs/api/supervised/trainers.html`). The last one, `.tools.shortcuts`, allows us to quickly build our network.

In this example, we will build a simple, single hidden layer network. Before we do this, however, we need to prepare the dataset:

```
def prepareANNDataset(data):
    '''
        Method to prepare the dataset for ANN training
        and testing
    '''
    # we only import this when preparing ANN dataset
    import pybrain.datasets as dt

    # supplementary method to convert list to tuple
```

```
def extract(row):
    return tuple(row)

# get the number of inputs and outputs
inputs = len(data[0].columns)
outputs = 2

# create dataset object
dataset = dt.SupervisedDataSet(inputs, outputs)

# convert dataframes to lists of tuples
x = list(data[0].apply(extract, axis=1))
y = [(item,abs(item - 1)) for item in data[1]]

# and add samples to the ANN dataset
for x_item, y_item in zip(x,y):
    dataset.addSample(x_item, y_item)

return dataset
```

We assume that input data is a tuple with two elements: the first is a DataFrame with all the independent variables and the second is a pandas Series structure with our dependent variable.

 You can think of a Series as a single column of a DataFrame.

First, in our method, we import `pybrain.datasets`. We do it this way so that we do not unnecessarily load this module into the memory when we do not need to use this method in our script.

We then determine how many inputs and outputs our network will have. The number of inputs is the number of columns in our input dataset. The number of outputs, as mentioned earlier, is the number of levels of our dependent variable. Using `.SupervisedDataSet(...)`, we create a skeleton of our ANN dataset. The x object holds all the input observations while the y object holds our target variables; these two structures are lists of tuples. To create x, we use the `extract(...)` method to convert our data (passed as a list) to a tuple; this is necessary to build a dataset to train our network. We use the `.apply(...)` method of DataFrame to apply the `extract(...)` method to each and every row in our DataFrame.

 The `extract(...)` method is accessible only to objects within the `prepareANNDataset(...)` method; you cannot use it from, say, `printModelSummary(...)`.

Our y object also holds a list of tuples. The tuples for y are created so they offset each other: if the first one is 0, then the other one is 1; we attain this using a simple math hack, (item, abs (item - 1)), where, if the client decided not to apply for a credit, that is, our target has a value of 0, we subtract 1 (getting -1) and take the absolute value of it (resulting in 1). We are essentially creating a true flag for a situation where a customer did not apply for a credit.

We went through all this hassle so that we can then use the .addSample (...) method to add the observations to our final dataset. The .addSample (...) method accepts tuples for the input and target variables.

Now that we have the dataset ready, we can train our network. Our fitANN (...) method takes the dataset and first determines the number of input and target neurons; we use the .shape attribute of the SupervisedDataSet input and target objects to get the number of columns.

Then, we create the actual ANN. We use a shortcut PyBrain has built-in: the .buildNetwork (...) method. The first unnamed parameter to the function is the number of neurons in the input layer, the second is the number of neurons in the hidden layer, and the third one, in our example, is the number of neurons in the output layer.

> The buidNetwork method can accept an arbitrary number of hidden layers. The method treats the last unnamed parameter as the number of neurons in the output layer.

We also specify the activation function for the hidden and output layers: the hiddenclass parameter specifies TanhLayer for the hidden layer and outclass specifies SoftmaxLayer. The tanh function squashes the inputs to fit between 0 and 1 with a similar-looking transfer function as the sigmoid.

> For the reasons that go beyond the scope of this book, tanh is a preferred activation function to the sigmoid one. See, for instance, the paper at http://yann.lecun.com/exdb/publis/pdf/lecun-98b.pdf.

The last parameter that we use is the bias. When set to true, it allows the summation function in soma to include a constant parameter that is also altered during training. Think about it in terms of a linear function: $y = AX + b$, where A is the vector of input weights, X is the vector of input variables, and b is the bias.

Now that we have our network defined, we need to specify the training mechanism. We will use the back-propagation algorithm to train our network. The `.BackpropTrainer(...)` method accepts our newly created network as its first parameter. The dataset that we created earlier is passed as a second parameter. We also specify two additional attributes: verbose so that we can track the progress of the training, and we switch off batch learning. Switching off the batch learning puts the training in an online mode; the online mode updates the weights and neurons' parameters after every observation. In contrast, the batch learning performs the updates to the structure of the network only once per training iteration (epoch).

The training iteration, or an epoch, is a period during which all the observations in our training dataset are presented to the network.

What is left now is to train our network. On the newly created trainer object, we invoke the `.trainUntilConvergence(...)` method.

You can also train the network for one epoch using the `.train()` method or for a number of epochs using `.trainEpochs(...)`. Check `http://pybrain.org/docs/api/supervised/trainers.html` for more details.

The method runs until it converges, that is, the next iteration brings no improvement to the training and/or validation dataset or if the `maxEpochs` number is reached. We also specify `validationProportion` to `0.25`, which means that we will be using a quarter of the training dataset to validate our model.

The validation dataset is a subset of the training dataset that is not used to train the network. The overarching aim of the ANN training is to minimize the error between the network's output and target variable. However, this may lead to a situation where the model fits each and every training observation perfectly (that is, the network's error reaches 0) but does not generalize well (see `https://clgiles.ist.psu.edu/papers/AAAI-97.overfitting.hard_to_do.pdf`). So, in order to avoid overfitting, the network tracks the error in the validation dataset; when the error in the validation dataset starts to rise, the training stops.

In our training scheme, we have set `continueEpochs` to `3` so if the trainer sees the error starting to rise in the validation dataset, it will continue for another three epochs before stopping. Doing so accounts for situations where the network finds a local minimum and, for another one or two epochs, the error raises before it starts to fall again.

Having trained the network, we can now predict the classes:

```
# classify the unseen data
predicted = classifier.activateOnDataset(testing)

# the lowest output activation gives the class
predicted = predicted.argmin(axis=1)
```

The `.activateOnDataset(...)` method takes the testing dataset and produces a prediction; for each observation in the testing dataset, the network is activated and produces a result. The predicted object now holds outputs that have two values; we want to find the index of the minimum as this is our class. We use the `.argmin(...)` method that returns just this.

ANNs take a bit of time to estimate as the structure is more complex than any other models presented earlier. In our example, the neural network performed on par with the SVM models presented earlier but took significantly longer to estimate:

```
The method fitANN took 113.17 sec to run.
Overall accuracy of the model is 91.10 percent
Classification report:
                precision    recall   f1-score    support

        0.0        0.93       0.97       0.95        11880
        1.0        0.67       0.44       0.53         1541

avg / total        0.90       0.91       0.90        13421

Confusion matrix:
 [[11551    329]
 [  865    676]]
ROC:  0.705491290801
```

Also, while you can readily analyze the coefficients of the models that we have seen so far, it is not so easy for ANNs. Unless it is a very simple network, the parameters of the network have no easy explanation. A neural network can sometimes be referred to as a black-box: a model that takes an input and produces a reasonable output but you have no means to assess how it does this.

I am not trying to imply that you should always stick with simpler models—my views are very far from it. Neural networks have proven hugely successful in areas where designing explicit models would simply be much more complex than designing and training an ANN. For instance, the models that aim at understating human speech or recognizing objects in photos in their explicit form would be extremely complex if one wanted to understand each and every component of the model and how it affects the output. If such an explicit knowledge is not required, ANNs normally do their job very well.

There's more...

With `PyBrain`, we can build more complex networks. In this example, we will build an ANN with two hidden layers:

```
# create the classifier object
ann = pb.buildNetwork(inputs_cnt,
        inputs_cnt * 2,
        inputs_cnt / 2,
        target_cnt,
        hiddenclass=st.TanhLayer,
        outclass=st.SoftmaxLayer,
        bias=True
    )
```

The network constructed has two hidden layers: the first one with 20 hidden neurons, and the second with five.

The investment to create and estimate a more complex model did not pay off—it took twice as long to estimate and the model performed worse than the simpler one:

```
The method fitANN took 769.27 sec to run.
Overall accuracy of the model is 91.21 percent
Classification report:
                precision    recall  f1-score   support

        0.0       0.94      0.96      0.95     12118
        1.0       0.64      0.50      0.56      1550

avg / total       0.90      0.91      0.91     13668

Confusion matrix:
 [[11688   430]
 [  771   779]]
ROC:  0.733548120897
```

See also

It goes far beyond the scope of this book to explain in detail the various structures of a neural network. In the introduction to this recipe, we tried to at least outline the structure so that you get a better understanding on how our model works. For mathematically adept readers interested in Artificial Neural Networks, I strongly recommend reading *Neural Networks and Learning Machines* by Simon O. Haykin, `http://www.amazon.com/Neural-Networks-Learning-Machines-Edition/dp/0131471392`.

4
Clustering Techniques

In this chapter, we will cover various techniques that will allow you to cluster the outbound call data of a bank that we used in the previous chapter. You will learn the following recipes:

▶ Assessing the performance of a clustering method

▶ Clustering data with the k-means algorithm

▶ Finding an optimal number of clusters for k-means

▶ Discovering clusters with the mean shift clustering model

▶ Building fuzzy clustering model with c-means

▶ Using a hierarchical model to cluster your data

▶ Finding groups of potential subscribers with DBSCAN and BIRCH algorithms

Introduction

Unlike a classification problem, where we know a class for each observation (often referred to as supervised training or training with a teacher), clustering models find patterns in data without requiring labels (called **unsupervised learning paradigm**).

The clustering methods put a set of unknown observations into buckets based on how similar the observations are. Such analysis aids the exploratory phase when an analyst wants to see if there are any patterns occurring naturally in the data.

Assessing the performance of a clustering method

Without knowing the true labels, we cannot use the metrics introduced in the previous chapter. In this recipe, we will introduce three measures that will help us assess the effectiveness of our clustering methods: Davis-Bouldin, Pseudo-F (sometimes referred to as Calinski-Harabasz), and Silhouette Score are internal evaluation metrics. In contrast, if we knew the true labels, we could use a range of measures, such as Adjusted Rand Index, Homogeneity, or Completeness scores, to name a few.

Refer to the documentation of `Scikit` on clustering methods for a deeper overview of various external evaluation metrics of clustering methods:

`http://scikit-learn.org/stable/modules/clustering.html#clustering-performance-evaluation`

For a list of internal clustering validation methods, refer to `http://datamining.rutgers.edu/publication/internalmeasures.pdf`.

Getting ready

To execute this recipe, you will need `pandas`, `NumPy`, and `Scikit`. No other prerequisites are required.

How to do it...

Out of the three internal evaluation metrics mentioned earlier, only Silhouette score is implemented in `Scikit`; the other two I developed for the purpose of this book. To assess your clustering model's performance, you can use the `printClusterSummary(...)` method from the `helper.py` file:

```
def printClustersSummary(data, labels, centroids):
    '''
        Helper method to automate model's assessment
    '''
    print('Pseudo_F: ', pseudo_F(data, labels, centroids))
    print('Davis-Bouldin: ',
        davis_bouldin(data, labels, centroids))
    print('Silhouette score: ',
        mt.silhouette_score(data, labels,
            metric='euclidean'))
```

How it works...

The first metric that we introduce is the Pseudo-F score. The maximum of Calinski-Harabasz heuristic over a number of models with the number of clusters indicates the one with the best data clustering.

 The Pseudo-F index is calculated as the ratio of the squared distance between the center of each cluster to the geometrical center of the whole dataset to the number of clusters minus one, multiplied by the number of observations in each cluster. This number is then divided by the ratio of squared distances between each point and the centroid of the cluster to the total number of observations less the number of clusters.

```
def pseudo_F(X, labels, centroids):
    mean = np.mean(X,axis=0)
    u, c = np.unique(labels, return_counts=True)

    B = np.sum([c[i] * ((clus - mean)**2)
            for i, clus in enumerate(centroids)])

    X = X.as_matrix()
    W = np.sum([(x - centroids[labels[i]])**2
                for i, x in enumerate(X)])

    k = len(centroids)
    n = len(X)

    return (B / (k-1)) / (W / (n-k))
```

First, we calculate the geometrical coordinates of the center of the whole dataset. Algebraically speaking, this is nothing more than the average of each column. Using `.unique(...)` of NumPy, we then count the number of observations in each cluster.

 When you pass `return_counts=True` to the `.unique(...)` method, it returns not only the list of unique values u in the labels vector, but also the counts c for each distinct value.

Next, we calculate the squared distances between the centroid of each cluster and center of our dataset. We create a list by using the `enumerate(...)` method through each element of our list of centroids and taking a square of the differences between each of the cluster and center, and multiplying it by the count of observations in that cluster: `c[i]`. The `.sum(...)` method of `NumPy`, as the name suggests, sums all the elements of this list. We then calculate the sum of the squared distances between each observation to the center of the cluster it belongs to. The method returns the Calinski-Harabasz Index.

In contrast to the Pseudo-F, the Davis-Bouldin metric measures the worst-case scenario of intercluster heterogeneity and intracluster homogeneity. Thus, the objective of finding the optimal number of clusters is to minimize this metric. Later in this chapter, in the *Finding an optimal number of clusters for k-means* recipe, we will develop a method that will find an optimal number of clusters by minimizing the Davis-Bouldin metric.

Check MathWorks (developers of Matlab) for the formula on how to calculate the Davis-Bouldin metric:

`http://au.mathworks.com/help/stats/clustering.`
`evaluation.daviesbouldinevaluation-class.html`

```python
def davis_bouldin(X, labels, centroids):
    distance = np.array([
        np.sqrt(np.sum((x - centroids[labels[i]])**2))
        for i, x in enumerate(X.as_matrix())])

    u, count = np.unique(labels, return_counts=True)

    Si = []

    for i, group in enumerate(u):
        Si.append(distance[labels == group].sum() / count[i])

    Mij = []

    for centroid in centroids:
        Mij.append([
            np.sqrt(np.sum((centroid - x)**2))
            for x in centroids])

    Rij = []
    for i in range(len(centroids)):
        Rij.append([
            0 if i == j
            else (Si[i] + Si[j]) / Mij[i][j]
            for j in range(len(centroids))])

    Di = [np.max(elem) for elem in Rij]

    return np.array(Di).sum() / len(centroids)
```

First, we calculate the geometrical distance between each observation and the centroid of the cluster it belongs to and count how many observations we have in each cluster.

`Si` measures the homogeneity within the cluster; effectively, it is an average distance between each observation in the cluster and its centroid. `Mij` quantifies the heterogeneity between clusters by calculating the geometrical distances between each clusters' centroids. `Rij` measures how well two clusters are separated and `Di` selects the worst-case scenario of such a separation. The Davis-Bouldin metric is an average of `Di`.

The Silhouette Score (Index) is another method of the internal evaluation of clusters. Even though we do not implement the method (as `Scikit` already provides the implementation), we will briefly describe how it is calculated and what it measures. A silhouette of a cluster measures the average distance between each and every observation in the cluster and relates it to the average distance between each point in each cluster. The metric can theoretically get any value between `-1` and `1`; the value of `-1` would mean that all the observations were inappropriately clustered in one cluster where, in fact, the more appropriate cluster would be one of the neighboring ones. Although theoretically possible in practice, you will most likely never encounter negative Silhouette Scores. On the other side of the spectrum, a value of `1` would mean a perfect separation of all the observations into appropriate clusters. A Silhouette Score value of `0` would mean a perfect overlap between clusters and, in practice, it would mean that every cluster would have an equal number of observations that should belong to other clusters.

See also...

For a more in-depth description of these (and many more algorithms), check `https://cran.r-project.org/web/packages/clusterCrit/vignettes/clusterCrit.pdf`.

Clustering data with k-means algorithm

The k-means clustering algorithm is likely the most widely known data mining technique for clustering vectorized data. It aims at partitioning the observations into discrete clusters based on the similarity between them; the deciding factor is the Euclidean distance between the observation and centroid of the nearest cluster.

Getting ready

To run this recipe, you need `pandas` and `Scikit`. No other prerequisites are required.

How to do it...

`Scikit` offers several clustering models in its cluster submodule. Here, we will use `.KMeans(...)` to estimate our clustering model (the `clustering_kmeans.py` file):

```
def findClusters_kmeans(data):
    '''
        Cluster data using k-means
```

```
'''
# create the classifier object
kmeans = cl.KMeans(
    n_clusters=4,
    n_jobs=-1,
    verbose=0,
    n_init=30
)

# fit the data
return kmeans.fit(data)
```

How it works...

Just like in the previous chapter (and in all the recipes that follow), we start with reading in the data and selecting the features that we want to discriminate our observations on:

```
# the file name of the dataset
r_filename = '../../Data/Chapter4/bank_contacts.csv'

# read the data
csv_read = pd.read_csv(r_filename)

# select variables
selected = csv_read[['n_duration','n_nr_employed',
        'prev_ctc_outcome_success','n_euribor3m',
        'n_cons_conf_idx','n_age','month_oct',
        'n_cons_price_idx','edu_university_degree','n_pdays',
        'dow_mon','job_student','job_technician',
        'job_housemaid','edu_basic_6y']]
```

We use the same dataset as in the previous chapter and limit the features to the most descriptive in our classification efforts. Also, we still use the @hlp.timeit decorator to measure how quickly our models estimate.

The .KMeans(...) method of Scikit accepts many options. The n_clusters parameter defines how many clusters to expect in the data and determines how many clusters the method will return. In the *Finding an optimal number of clusters for k-means* recipe, we will develop an iterative method of finding the optimal number of clusters for the k-means clustering algorithm.

The n_jobs parameter is the number of jobs to be run in parallel by your machine; specifying -1 instructs the method to spin off as many parallel jobs as the number of cores that the processor of your machine has. You can also specify the number of jobs explicitly by passing an integer, for example, 8.

The `verbose` parameter controls how much you will see about the estimation phase; setting this parameter to 1 will print out the details about the estimation.

The `n_init` parameter controls how many models to estimate. Every run of a k-means algorithm starts with randomly selecting the centroids of each cluster and then iteratively refining these to arrive at a model with the best intercluster separation and intracluster similarity. The `.KMeans(...)` methods builds as many models as specified by the `n_init` parameter with varying starting conditions (randomized initial seeds for the centers) and then selects the one that performed the best in terms of inertia.

 Inertia measures the intracluster variation. It is a within-cluster sum of squares. We can also talk about **explained inertia** that would be a ratio of the intracluster sum of squares to the total sum of squares.

There's more...

Of course, `Scikit` is not the only way to estimate a k-means clustering model; we can also use `SciPy` to do this (the `clustering_kmeans_alternative.py` file):

```
import scipy.cluster.vq as vq

def findClusters_kmeans(data):
    '''
        Cluster data using k-means
    '''
    # whiten the observations
    data_w = vq.whiten(data)

    # create the classifier object
    kmeans, labels = vq.kmeans2(
        data_w,
        k=4,
        iter=30
    )

    # fit the data
    return kmeans, labels
```

In contrast to the way we did it with `Scikit`, we need to whiten the data first with `SciPy`. Whitening is similar to standardizing the data (refer to the *Normalizing and standardizing the features* recipe in *Chapter 1, Preparing the Data*) with the exception that it does not remove the mean; `.whiten(...)` unifies the variance to 1 across all the features.

The first parameter passed to `.kmeans2(...)` is the whitened dataset. The `k` parameter specifies how many clusters to fit the data into. In our example, we allow the `.kmeans2(...)` method to run 30 iterations at most (the `iter` parameter); if the method does not converge by then, it will stop and return the estimates from the 30th iteration.

See also...

▶ `MLPy` also provides a method to estimate the k-means model: `http://mlpy.sourceforge.net/docs/3.5/cluster.html#k-means`

Finding an optimal number of clusters for k-means

Often, you will not know how many clusters you can expect in your data. For two or three-dimensional data, you could plot the dataset in an attempt to eyeball the clusters. However, it becomes harder with a dataset that has many dimensions as, beyond three dimensions, it is impossible to plot the data on one chart.

In this recipe, we will show you how to find the optimal number of clusters for a k-means clustering model. We will be using the Davis-Bouldin metric to assess the performance of our k-means models when we vary the number of clusters. The aim is to stop when a minimum of the metric is found.

Getting ready

In order to execute this, you will need `pandas`, `NumPy`, and `Scikit`. No other prerequisites are required.

How to do it...

In order to find the optimal number of clusters, we developed the `findOptimalClusterNumber(...)` method. The overall algorithm of estimating the k-means model has not changed—instead of calling `findClusters_kmeans(...)` (that we defined in the first recipe of this chapter), we first call `findOptimalClusterNumber(...)` (the `clustering_kmeans_search.py` file):

```
optimal_n_clusters = findOptimalClusterNumber(selected)
```

The `findOptimalClusterNumber(...)` is defined as follows:

```
def findOptimalClusterNumber(
        data,
        keep_going = 1,
```

```
        max_iter = 30
):
'''
    A method that iteratively searches for the
    number of clusters that minimizes the Davis-Bouldin
    criterion
'''
# the object to hold measures
measures = [666]

# starting point
n_clusters = 2

# counter for the number of iterations past the local
# minimum
keep_going_cnt = 0
stop = False    # flag for the algorithm stop

# main loop
# loop until minimum found or maximum iterations reached
while not stop and n_clusters < (max_iter + 2):
    # cluster the data
    cluster = findClusters_kmeans(data, n_clusters)

    # assess the clusters effectiveness
    labels = cluster.labels_
    centroids = cluster.cluster_centers_

    # store the measures
    measures.append(
        hlp.davis_bouldin(data, labels, centroids)
    )

    # check if minimum found
    stop = checkMinimum(keep_going)

    # increase the iteration
    n_clusters += 1

# once found -- return the index of the minimum
return measures.index(np.min(measures)) + 1
```

How it works...

To use `findOptimalClusterNumber(...)`, you need to pass at least the dataset as the first parameter: obviously, without data, we cannot estimate any models. The `keep_going` parameter defines how many iterations to continue if the current model's Davis-Bouldin metric has a greater value than the minimum of all the previous iterations. This way, we are able to mitigate (to some extent) the problem of stopping at one of the local minima and not reaching the global one:

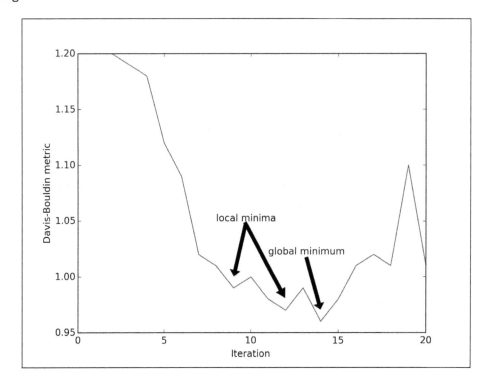

The `max_iter` parameter specifies the maximum number of models to build; our method starts building k-means models with `n_clusters = 2` and then iteratively continues until either the global minimum is found or we reach the maximum number of iterations.

Our `findOptimalClusterNumber(...)` method starts with defining the parameters of the run.

The measures list will be used to store the subsequent values of the Davis-Bouldin metrics for the estimated models; as our aim is to find the minimum of the Davis-Bouldin metric, we select an arbitrarily large number as the first element so that it does not become our minimum.

The `n_clusters` method defines the starting point: the first k-means model will aim at clustering the data into two buckets.

The `keep_going_cnt` is used to track how many iterations to continue past the most recent value of the Davis-Bouldin that was higher than the previously found minimum. With the preceding image presented as an example, our method would not stop at any of the local minima; even though the Davis-Bouldin metric for iterations 10 and 13 were greater than the minima attained at iterations 9 and 12 (respectively), the models with 11 and 14 clusters showed lower values. In this example, we stop at iteration 14 as models with 15 and 16 clusters show a greater value of the Davis-Bouldin metric.

The stop flag is used to control the execution of the main loop of our method. Our while loop continues until we find that either the minimum or maximum number of iterations has been reached.

The beauty of Python's syntax can clearly be seen when one reads the code while not stop and `n_clusters < (max_iter + 2)` and can easily translate it to English. This, in my opinion, makes the code much more readable and maintainable as well as less error-prone.

The loop starts with estimating the model with a defined number of clusters:

```
cluster = findClusters_kmeans(data, n_clusters)
```

Once the model is estimated, we get the estimated labels and centroids and calculate the Davis-Bouldin metric using the `.davis_bouldin(...)` method we presented in the first recipe for this chapter. The metric is appended using the `append(...)` method to the measures list.

Now, we need to check whether the newly estimated model has the Davis-Bouldin metric smaller than all the previously estimated models. For this, we use the `checkMinimum(...)` method. The method can only be accessed by the `findOptimalClusterNumber(...)` method:

```
def checkMinimum(keep_going):
    '''
        A method to check if minimum found
    '''
    global keep_going_cnt # access global counter

    # if the new measure is greater than for one of the
    # previous runs
    if measures[-1] > np.min(measures[:-1]):
        # increase the counter
        keep_going_cnt += 1

        # if the counter is bigger than allowed
        if keep_going_cnt > keep_going:
            # the minimum is found
```

```
        return True
# else, reset the counter and return False
else:
    keep_going_cnt = 0

return False
```

First, we define `keep_going_cnt` as a global variable. By doing so, we do not need to pass the variable to `checkMinimum(...)` and we can keep track of the counter globally. In my opinion, it aids the readability of the code.

Next, we compare the results of the most recently estimated model with the minimum already found in previous runs; we use the `.min(...)` method of `NumPy` to get the minimum. If the current run's Davis-Bouldin metric is greater than the minimum previously found, we increase the `keep_going_cnt` counter by one and check whether we have exceeded the number of `keep_going` iterations—if the number of iterations is exceeded the method returns `True`. Returning `True` means that we have found the global minimum (given the current assumptions specified by `keep_going`). If, however, the newly estimated model has a lower value of the Davis-Bouldin metric, we first reset the `keep_going_cnt` counter to 0 and return `False`. We also return `False` if we did not exceed the `keep_going` specified number of iterations.

Now that we know whether we have or have not found the global minimum, we increase n_clusters by 1.

The loop continues until the `checkMinimum(...)` method returns `True` or n_clusters exceeds `max_iter + 2`. We increase `max_iter` by 2 as we start our algorithm by estimating the k-means model with two clusters.

Once we break out of our while loop, we return the index of the minimum found increased by 1. The increment is necessary as the index of the Davis-Bouldin metric for the model with n_cluster = 2 is 1.

 The list indexing starts with 0.

Now we can estimate the model with the optimal number of clusters and print out its metrics:

```
# cluster the data
cluster = findClusters_kmeans(selected, optimal_n_clusters)

# assess the clusters effectiveness
labels = cluster.labels_
centroids = cluster.cluster_centers_

hlp.printClustersSummary(selected, labels, centroids)
```

There's more...

Even though I mentioned earlier that it is not so easy to spot clusters of data, a pair-plot might be sometimes useful to explore the dataset visually (the `clustering_kmeans_search_alternative.py` file):

```python
import seaborn as sns
import matplotlib.pyplot as plt

def plotInteractions(data, n_clusters):
    '''
        Plot the interactions between variables
    '''
    # cluster the data
    cluster = findClusters_kmeans(data, n_clusters)

    # append the labels to the dataset for ease of plotting
    data['clus'] = cluster.labels_

    # prepare the plot
    ax = sns.pairplot(selected, hue='clus')

    # and save the figure
    ax.savefig(
        '../../Data/Chapter4/k_means_{0}_clusters.png' \
        .format(n_clusters)
    )
```

We use `Seaborn` and `Matplotlib` to plot the interactions. First, we estimate the model with predefined `n_clusters`. Then, we append the clusters to the main dataset. Using the `.pairplot(...)` method of `Seaborn`, we create a chart that iterates through each combination of features and plots the resulting interactions. Lastly, we save the chart.

The result (limited for three features for readability) with 14 clusters should look similar to the following one:

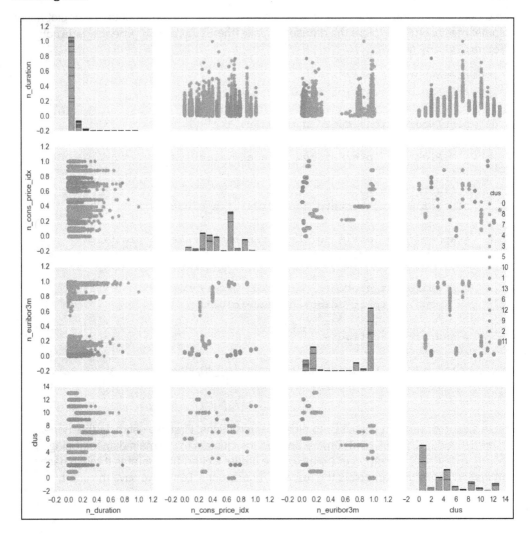

Even with a small number of dimensions, it is really hard to *eyeball* how many clusters there should be in this dataset. For other data, it might be beneficial to actually create a chart like this. The problem with this method is especially visible when you deal with discrete or categorical variables.

Discovering clusters with mean shift clustering model

A method similar in terms of finding centers (or maxima of density) is the Mean Shift model. In contrast to the k-means, the method does not require specifying the number of clusters—the model returns the number of clusters based on the number of density centers found in the data.

Getting ready

To estimate this model, you will need `pandas` and `Scikit`. No other prerequisites are required.

How to do it...

We start the estimation in a similar way as with the previous models—by reading the dataset in and limiting the number of features. Then, we use `findClusters_meanShift(...)` to estimate the model (the `clustering_meanShift.py` file):

```
def findClusters_meanShift(data):
    '''
        Cluster data using Mean Shift method
    '''
    bandwidth = cl.estimate_bandwidth(data,
        quantile=0.25, n_samples=500)

    # create the classifier object
    meanShift = cl.MeanShift(
        bandwidth=bandwidth,
        bin_seeding=True
    )

    # fit the data
    return meanShift.fit(data)
```

How it works...

First, we `estimate_bandwidth(...)`. The bandwidth is used in the RBF kernel of the method.

 We used Radial Basis Functions previously in the SVM classification recipe: check the *Utilizing Support Vector Machines as a classification engine* recipe in *Chapter 3, Classification Techniques*.

The `estimate_bandwidth(...)` method takes data as its parameter at least.

For samples with many observations, as the algorithm used in the method grows quadratically with the number of observations, it is wise to limit the number of records; use `n_samples` to do this.

The `quantile` parameter determines where to cut off the samples passed to the kernel in the `.MeanShift(...)` method.

 The `quantile` value of `0.5` specifies the median.

Now, we are ready to estimate the model. We build the model object by passing the `bandwidth` and `bin_seeding` (optional) parameters.

If the `bin_seeding` parameter is set to `True`, the initial kernel locations are set to discretized groups (using the `bandwidth` parameter) of all the data points. Setting this parameter to `True` speeds up the algorithm as fewer kernel seeds will be initialized; with the number of observations in hundreds of thousands, this might translate to a significant speed up.

While we talk about the speed of estimation, this method expectedly performs much slower than any of the k-means (even with tens of clusters). On average, it took around 13 seconds on our machine to estimate: between 5 to 20 times slower than k-means.

See also...

▶ For those who are interested in reading about the innards of the Mean Shift algorithm, I recommend the following document: `http://homepages.inf.ed.ac.uk/rbf/CVonline/LOCAL_COPIES/TUZEL1/MeanShift.pdf`

▶ Refer to *Utilizing Support Vector Machines as a classification engine* from *Chapter 3, Classification Techniques*

Building fuzzy clustering model with c-means

K-means and Mean Shift clustering algorithms put observations into distinct clusters: an observation can belong to one and only one cluster of similar samples. While this might be right for discretely separable datasets, if some of the data overlaps, it may be too hard to place them into only one bucket. After all, our world is not just black or white but our eyes can register millions of colors.

The c-means clustering model allows each and every observation to be a member of more than one cluster and this membership is weighted: the sum of all the weights across all the clusters for each observation must equal 1.

Getting ready

To execute this recipe, you will need `pandas` and the `Scikit-Fuzzy` module. The `Scikit-Fuzzy` module normally does not come preinstalled with Anaconda so you will need to install it yourself.

In order to do so, clone the `Scikit-Fuzzy` repository to a local folder:

git clone https://github.com/scikit-fuzzy/scikit-fuzzy.git

On finishing the cloning, `cd` into the `scikit-fuzzy-master/` folder and execute the following command:

python setup.py install

This should install the module.

How to do it...

The algorithm to estimate c-means is provided with `Scikit`. We call the method slightly differently from the already presented methods (the `clustering_cmeans.py` file):

```python
import skfuzzy.cluster as cl
import numpy as np

def findClusters_cmeans(data):
    '''
        Cluster data using fuzzy c-means clustering
        algorithm
    '''
    # create the classifier object
    return cl.cmeans(
        data,
        c = 5,              # number of clusters
        m = 2,              # exponentiation factor

        # stopping criteria
        error = 0.01,
        maxiter = 300
    )

# cluster the data
centroids, u, u0, d, jm, p, fpc = findClusters_cmeans(
    selected.transpose()
)
```

How it works...

As with the previous recipes, we first load the data and select the columns that we want to use to estimate the model.

Then, we call the `findClusters_cmeans(...)` method to estimate the model. In contrast to all the previous methods, where we first created an untrained model and then fit the data using the `.fit(...)` method, `.cmeans(...)` estimates the model upon creation using data passed to it (along with all the other parameters).

The `c` parameter specifies how many clusters to fit while the `m` parameter specifies the exponentiation factor applied to the membership function during each iteration.

> Be careful with how many clusters you are trying to find as specifying too many will result in errors when calculating the metrics as some of the clusters may have 0 members! If you are getting `IndexError` and the estimation stops, reduce the number of clusters without the loss of accuracy.

We also specify the stopping criteria. Our model will stop iterating if the difference between the current and previous iterations (in terms of the change in the membership function) is smaller than `0.01`.

The model returns a number of objects. The centroids holds the coordinates of all the five clusters. The `u` holds the values of membership for each and every observation; it has the following structure:

```
The method findClusters_cmeans took 0.93 sec to run.
[[ 0.15019766  0.05824843  0.04623635 ...,  0.14150561  0.26927404
   0.14128503]
 [ 0.13702982  0.05074458  0.0402064  ...,  0.28432347  0.27960814
   0.38820845]
 [ 0.37076827  0.74075993  0.79335671 ...,  0.15009361  0.14779614
   0.14957908]
 [ 0.14041724  0.05272835  0.04176752 ...,  0.2576644   0.13334643
   0.13312653]
 [ 0.20158702  0.0975187   0.07843302 ...,  0.16641291  0.16997526
   0.18780091]]
Pseudo_F:  8340.93964306
Davis-Bouldin:  1.30629514194
```

If you sum up the values in each column across all the rows, you will find that each time it sums up to `1` (as expected).

The remaining returned objects are of less interest to us: `u0` are the initial membership seeds for each observation, `d` is the final Euclidean distance matrix, `jm` is the history of changes of the objective function, `p` is the number of iterations it took to estimate the model, and `fpc` is the fuzzy partition coefficient.

In order to be able to calculate our metrics, we need to assign the cluster to each observation. We achieve this with the .argmax(...) function of NumPy:

```
labels = [
    np.argmax(elem) for elem in u.transpose()
]
```

The method returns the index of the maximum element from the list. As the initial layout of our u matrix was *n_cluster x n_sample*, we first need to .transpose() the u matrix so that we can iterate through rows where each row is an observation from our sample.

Using hierarchical model to cluster your data

The hierarchical clustering model aims at building a hierarchy of clusters. Conceptually, you might think of it as a decision tree of clusters: based on the similarity (or dissimilarity) between clusters, they are aggregated (or divided) into more general (more specific) clusters. The agglomerative approach is often referred to as **bottom up**, while the divisive is called **top down**.

Getting ready

To execute this recipe, you will need pandas, SciPy, and PyLab. No other prerequisites are required.

How to do it...

Hierarchical clustering can be extremely slow for big datasets as the complexity of the agglomerative algorithm is $O(n^3)$. To estimate our model, we use a single-linkage algorithm that has better complexity, $O(n^2)$, but can still be very slow for large datasets (the clustering_hierarchical.py file):

```
def findClusters_link(data):
    '''
        Cluster data using single linkage hierarchical
        clustering
    '''
    # return the linkage object
    return cl.linkage(data, method='single')
```

How it works...

The code is extremely simple: all we do is call the .linkage(...) method of SciPy, pass the data, and specify the method.

Under the hood, `.linkage(...)`, based on the method chosen, aggregates (or splits) the clusters based on a specific distance metric; `method='single'` aggregates the clusters based on the minimum distance between each and every point in two clusters being considered (hence, the $O(n^2)$).

For a list of all the metrics, check the SciPy documentation:
`http://docs.scipy.org/doc/scipy/reference/`
`generated/scipy.cluster.hierarchy.linkage.`
`html#scipy.cluster.hierarchy.linkage`

We can visualize the aggregations (or splits) in a form of a tree:

```
import pylab as pl

# cluster the data
cluster = findClusters_ward(selected)

# plot the clusters
fig  = pl.figure(figsize=(16,9))
ax   = fig.add_axes([0.1, 0.1, 0.8, 0.8])
dend = cl.dendrogram(cluster, truncate_mode='level', p=20)
ax.set_xticks([])
ax.set_yticks([])

fig.savefig(
    '../../Data/Chapter4/hierarchical_dendrogram.png',
    dpi=300
)
```

First, we create a figure using PyLab and `.add_axes(...)`. The list passed to the function is the coordinates of the plot: `[x, y, width, height]` normalized to be between 0 and 1 where 1 means 100%. The `.dendrogram(...)` method takes the list of all the linkages, cluster, and creates the plot. We do not want to output all the possible links so we truncate the tree at the `p=20` level. We also do not need x or y ticks so we switch these off using the `.set_{x,y}ticks(...)` method.

For our dataset, your tree should look similar to the following one:

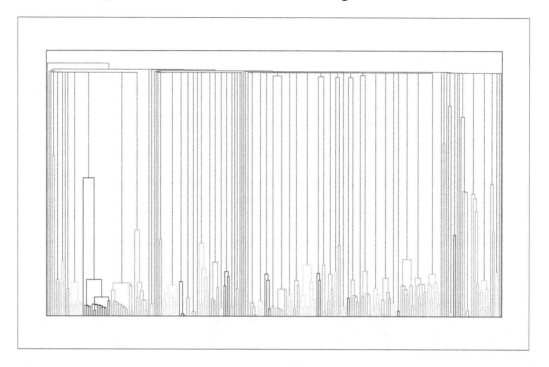

You can see that the closer you get to the top, the less distinct clusters you get (more aggregation occurs).

There's more...

We can also use MLPy to estimate the hierarchical clustering model (the clustering_hierarchical_alternative.py file):

```python
import mlpy as ml
import numpy as np

def findClusters_ward(data):
    '''
        Cluster data using Ward's hierarchical clustering
    '''
    # create the classifier object
    ward = ml.MFastHCluster(
        method='ward'
    )

    # fit the data
```

```
        ward.linkage(data)

        return ward

    # cluster the data
    cluster = findClusters_ward(selected)

    # assess the clusters effectiveness
    labels = cluster.cut(20)
    centroids = hlp.getCentroids(selected, labels)
```

We are using the `.MFastHCluster(...)` method with the `ward` method parameter. The Ward method of hierarchical clustering uses the Ward variance minimization algorithm. The `.linkage(...)` method can be thought of as an equivalent of `.fit(...)` for k-means; it fits the model to our data and finds all the linkages.

Once estimated, we use the `.cut(...)` method to trim the tree at level 20 so that the resulting number of clusters is manageable. The returned object contains the cluster membership index for each observation in the dataset.

As the Ward clustering method does not return centroids of the clusters (as these would vary depending at what level we cut the tree), in order to use our performance assessment methods, we created a new method that returns the centroids given the labels:

```
import numpy as np

def getCentroids(data, labels):
    '''
        Method to get the centroids of clusters in clustering
        models that do not return the centroids explicitly
    '''
    # create a copy of the data
    data = data.copy()

    # apply labels
    data['predicted'] = labels

    # and return the centroids
    return np.array(data.groupby('predicted').agg('mean'))
```

First, we create a local copy of the data (as we do not want to modify the object that was passed to the method) and create a new column called `predicted`. We then calculate the mean for each of the distinct levels of the `predicted` column and cast it as `np.array(...)`.

Now that we have centroids, we can assess the performance of the model.

See also...

If you are more interested in hierarchical clustering, I suggest reading the following text: http://www.econ.upf.edu/~michael/stanford/maeb7.pdf.

Finding groups of potential subscribers with DBSCAN and BIRCH algorithms

Density-based Spatial Clustering of Applications with Noise (DBSCAN) and **Balanced Iterative Reducing and Clustering using Hierarchies (BIRCH)** algorithms were the first approaches developed to handle noisy data effectively. Noise here is understood as data points that seem completely out of place when compared with the rest of the dataset; DBSCAN puts such observations into an unclassified bucket while BIRCH treats them as outliers and removes them from the dataset.

Getting ready

To execute this recipe, you will need `pandas` and `Scikit`. No other prerequisites are required.

How to do it...

Both the algorithms can be found in `Scikit`. To use `DBSCAN`, use the code found in the `clustering_dbscan.py` file:

```
import sklearn.cluster as cl

def findClusters_DBSCAN(data):
    '''
        Cluster data using DBSCAN algorithm
    '''
    # create the classifier object
    dbscan = cl.DBSCAN(eps=1.2, min_samples=200)

    # fit the data
    return dbscan.fit(data)

# cluster the data
cluster = findClusters_DBSCAN(selected)

# assess the clusters effectiveness
labels = cluster.labels_ + 1
centroids = hlp.getCentroids(selected, labels)
```

For the BIRCH model, check the script found in `clustering_birch.py`:

```python
import sklearn.cluster as cl

def findClusters_Birch(data):
    '''
        Cluster data using BIRCH algorithm
    '''
    # create the classifier object
    birch = cl.Birch(
        branching_factor=100,
        n_clusters=4,
        compute_labels=True,
        copy=True
    )

    # fit the data
    return birch.fit(data)

# cluster the data
cluster = findClusters_Birch(selected)

# assess the clusters effectiveness
labels = cluster.labels_
centroids = hlp.getCentroids(selected, labels)
```

How it works...

To estimate DBSCAN, we use the `.DBSCAN(...)` method of `Scikit`. This method can accept a number of different parameters. The eps controls the maximum distance between two samples to be considered in the same neighborhood. The `min_samples` parameter controls the neighborhood; once this threshold is exceeded, the point with at least that many neighbors is considered to be a core point.

As mentioned earlier, the leftovers (outliers) in DBSCAN are put into an unclassified bucket, which is returned with a -1 label. Before we can assess the performance of our clustering method, we need to shift the labels so that they start at 0 (with 0 being the unclustered data points). The method does not return the coordinates of the centroids so we calculate these ourselves using the `.getCentroids(...)` method introduced in the previous recipe.

Estimating BIRCH is equally effortless (on our part!). We use the `.Birch(...)` method of `Scikit`. The `branching_factor` parameter controls how many subclusters (or observations) to hold at maximum in the parent node of the tree; if the number is exceeded, the clusters (and subclusters) are split recursively. The `n_clusters` parameter instructs the method on how many clusters we would like the data to be clustered into. Setting `compute_labels` to `True` tells the method, on finishing, to prepare and store the label for each observation.

The outliers in the BIRCH algorithm are discarded. If you prefer the method to not alter your original dataset, you can set the `copy` parameter to `True`: this will copy the original data.

See also...

Here is a link to the original paper that introduced DBSCAN:

```
https://www.aaai.org/Papers/KDD/1996/KDD96-037.pdf
```

The paper that introduced BIRCH can be found here:

```
http://www.cs.sfu.ca/CourseCentral/459/han/papers/zhang96.pdf
```

I also suggest reading the documentation of the `.DBSCAN(...)` and `.Birch(...)` methods:

```
http://scikit-learn.org/stable/modules/generated/sklearn.cluster.
DBSCAN.html#sklearn.cluster.DBSCAN
```

```
http://scikit-learn.org/stable/modules/generated/sklearn.cluster.
Birch.html#sklearn.cluster.Birch
```

5
Reducing Dimensions

In this chapter, we will cover various techniques to reduce dimensions of your data. You will learn the following recipes:

- ▶ Creating three-dimensional scatter plots to present principal components
- ▶ Reducing the dimensions using the kernel version of PCA
- ▶ Using Principal Component Analysis to find things that matter
- ▶ Finding the principal components in your data using randomized PCA
- ▶ Extracting the useful dimensions using Linear Discriminant Analysis
- ▶ Using various dimension reduction techniques to classify calls using the k-Nearest Neighbors classification model

Introduction

The abundance of data available nowadays can be mind-boggling; the datasets grow not only in terms of the number of observations, but also get richer in terms of collected metadata.

In this chapter, we will present techniques that will allow you to extract the most important features from your data and use them in modeling. The drawback of using principal components instead of raw features in modeling is that it is almost impossible to meaningfully explain the models' coefficients, that is, understand the causality or what drives your predictions or classifications.

If your aim is to have a forecaster with the highest attainable accuracy and the focus of your project is not to understand the drivers, some of the following methods presented might be of interest to you.

Creating three-dimensional scatter plots to present principal components

Principal components are nothing more than multidimensional vectors that we can use to transform our data. By finding the principal dimensions of our data, we can discover a different picture that our data paints.

Getting ready

To execute this recipe, you will need `Matplotlib` with MPL Toolkits. These should come preinstalled if you are using the Anaconda distribution of Python. No other prerequisites are required.

How to do it...

To plot our three-dimensional data, we will use the `plot_components(...)` method (the `helper.py` file):

```python
def plot_components(z, y, color_marker, **f_params):
    '''
        Produce and save the chart presenting 3 principal
        components
    '''
    # import necessary modules
    import matplotlib.pyplot as plt
    from mpl_toolkits.mplot3d import Axes3D

    # convert the dependent into a Numpy array
    # this is done so z and y are in the same format
    y_np = y

    # do it only, however, if y is not NumPy array
    if type(y_np) != np.array:
        y_np = np.array(y_np)

    # create a figure
    fig = plt.figure()
    ax = fig.add_subplot(111, projection='3d')

    # plot the dots
    for i, j in enumerate(np.unique(y_np)):
        ax.scatter(
            z[y_np == j, 0],
            z[y_np == j, 1],
```

```
            z[y_np == j, 2],
            c=color_marker[i][0],
            marker=color_marker[i][1])

    ax.set_xlabel('First component')
    ax.set_ylabel('Second component')
    ax.set_zlabel('Third component')

    # save the figure
    plt.savefig(**f_params)
```

How it works...

The `plot_components(...)` method accepts a multitude of parameters. The z parameter is our dataset transformed into a three-dimensional space (that is, with only the top three principal components). The y parameter is a vector of our classes: we are using the same dataset as in the previous two chapters (with a limited number of features), so this is an indicator whether the call made by the banks employee resulted in a customer signing up for a credit or not. The `color_marker` is a list of tuples of the (color, marker) footprint, where each element holds the color indicator in the first position and the marker indicator in the second, for example, (`'red'`, `'o'`); in this example, we would be plotting small circles in a red color.

The last parameter is something that we have only touched upon in this book: `**f_params`. The `**f_params` parameter (or more commonly `**kwargs`) allows you to pass a variable number of keyword parameters to a function. For example, say that you have a method:

```
def print_person(**person_details):
    if person_details is not None:
        print('\nPerson details: ')

        for key in person_details:
            print(key + ': ' + person_details[key])
```

Now, calling the method as follows would produce a different output:

```
print_person(name='Jane', lastName='Doe')
print_person(nick='Missy')
Person details:
name: Jane
lastName: Doe


Person details:
nick: Missy
```

You can also pass a variable number of non-keyword parameters to a function using the $*args$ parameter.

In our `plot_components(...)` method, we first import the necessary modules: pyplot and Axes3D from mplot3d; the latter allows us to produce three-dimensional plots. Then, we convert our y vector of labels to a `NumPy` array, if it is not yet a `NumPy` array.

Next, we start plotting. First, we create a figure and `add_subplot(...)`. 111 means that you have to add the plot to the first row (first 1) and first column (second 1) and it's the first and only plot of this chart (third 1).

Passing 211 would add a subplot (to a figure of 2 subplots) in the first row and first column, while 212 would add the second subplot (the last 2) to the same chart. In other words, the number in the first position determines how many rows there are in the grid of the chart, the number in the second position defines the number of columns, while the last number indicates the cell to put the subplot in; the cells are numbered left to right and top to bottom.

The projection parameter instructs the `add_subplot(...)` method that we will be passing data in three dimensions.

The for loop iterates through a unique list of the classes that we have in y and adds scatter plots to the plot. For each class, we select our three principal components' values; you should already be familiar with the `z[y_np == j, 0]` notation, where we select only those elements of z for which the element at the same index in y_np is equal to our class label j, and we select the first column (the 0th index of the array).

The c parameter of scatter defines the color used to plot the points and marker determines what kind of markers we can use.

Check `http://matplotlib.org/api/markers_api.html` for the list of all the markers you can use to differentiate the data points on your charts.

The `set_{x,y,z}label(...)` method, as the name suggests, will put descriptions to the axes of our chart.

Lastly, we save the plot using the `.savefig(...)` method of `Matplotlib`. At a minimum, we need to pass the filename parameter. As we will be saving in a PNG format, we will also be passing the dpi (Dots Per Inch) parameter of 300, so, should you require, you can print the chart; for a web presentation, a 100 dpi should be enough (some use 72 dpi).

The resulting chart might look as follows:

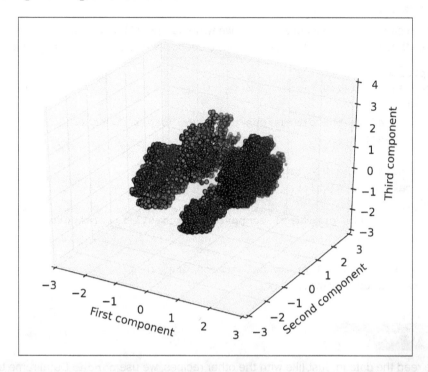

Reducing the dimensions using the kernel version of PCA

Principal Components Analysis (**PCA**) transforms a correlated set of variables into a set of principal components: variables that are linearly uncorrelated (orthogonal). PCA can produce as many principal components as there are variables but normally it would reduce the dimensionality of your data. The first principal component accounts for the highest amount of variability in the data, with the following principal components accounting for decreasingly less variance explained and the restriction of orthogonality (uncorrelated) to the other principal components.

Getting ready

To execute this recipe, you will need pandas, NumPy, and MLPY. For the plotting, you will need Matplotlib with MPL Toolkits. No other prerequisites are required.

How to do it...

In a fashion similar to the previous recipes, we wrap our model building efforts within a method so that we can time it using the `timeit` decorator (the `reduce_pca.py` file):

```
@hlp.timeit
def reduce_PCA(x):
    '''
        Reduce the dimensions using Principal Component
        Analysis
    '''
    # create the PCA object
    pca = ml.PCA(whiten=True)

    # learn the principal components from all the features
    pca.learn(x)

    # return only 3 principal components
    return pca.transform(x, k=3)
```

How it works...

First, we read the data in. Just like with the other recipes, we use `pandas` DataFrame to hold our data:

```
# the file name of the dataset
r_filename = '../../Data/Chapter5/bank_contacts.csv'

# read the data
csv_read = pd.read_csv(r_filename)
```

We then extract our dependent variable and independent variables from the `csv_read` object as we only use the independent variables in PCA:

```
x = csv_read[csv_read.columns[:-1]]
y = csv_read[csv_read.columns[-1]]
```

Finally, we call the `reduce_PCA(...)` method. The method requires only one parameter, x:

```
z = reduce_PCA(x)
```

In the method, we first create the PCA(...) object. The whiten parameter set to True causes the PCA(...) object to normalize the data so that each feature has a standard deviation of 1.

 We introduced whitening in the previous chapter. See *Clustering data with k-means algorithm* recipe in *Chapter 4, Clustering Techniques*.

You could also specify what method the PCA(...) algorithm implemented in MLPY should use: either 'svd' or 'cov'.

 For those of you who are more mathematically inclined and curious about the innards of PCA (and its relationship with Singular Value Decomposition), I recommend reading this paper:

https://www.cs.princeton.edu/picasso/mats/PCA-Tutorial-Intuition_jp.pdf

Next, we instruct the pca object to .learn(...) the principal components from the data x. On my computer, this is really fast and normally takes around 0.26 seconds (for this dataset).

 Note, of course, that the bigger your dataset (more rows and columns), the slower it will get. Should that happen, MLPY has another method to find the principal components: FastPCA(...). Check http://mlpy.sourceforge.net/docs/3.5/dim_red.html#fast-principal-component-analysis-pcafast for more. We will also show the Randomized PCA method later in this chapter.

Finally, we return pca and store it in the z object.

Now, it is time to plot the components:

```
# plot and save the chart
# to vary the colors and markers for the points
color_marker = [('r','o'),('g','.')]

file_save_params = {
    'filename': '../../Data/Chapter5/charts/pca_3d.png',
    'dpi': 300
}

hlp.plot_components(z.transform(x, k=3), y,
    color_marker, **file_save_params)
```

The `color_marker` list holds the definitions of our markers: if the call resulted in no credit sign up, we will plot small red circles, and if it did, then green dots. The `file_save_params` method holds all the keyword parameters for the `.savefig(...)` method that we call in the `.plot_components(...)` method.

Finally, we call the `.plot_components(...)` method. The `.transform(...)` method returns a list of the first three principal components (the `k=3` parameter).

There's more...

Scikit also has a method to find the principal components in your data:

```
@hlp.timeit
def reduce_PCA(x):
    '''
        Reduce the dimensions using Principal Component
        Analysis
    '''
    # create the PCA object
    pca = dc.PCA(n_components=3, whiten=True)

    # learn the principal components from all the features
    return pca.fit(x)
```

As usual, we first create the `pca` object and fit the data to it, that is, learn the principal components. In contrast to `MLPY`, the `.PCA(...)` method of `Scikit` allows you to specify the number of components that you want to find (or have returned from the method); not specifying this parameter would find all the possible principal components (up to the number of features in your dataset).

Another difference from `MLPY` is that `Scikit`'s `PCA(...)` method has an indicator of how much variance in the data each specific principal component accounts for. We can present it as follows:

```
# how much variance each component explains?
print(z.explained_variance_ratio_)

# and total variance accounted for
print(np.sum(z.explained_variance_ratio_))
```

The components found are slightly differently distributed than the ones found by the MLPY's
PCA(...) method:

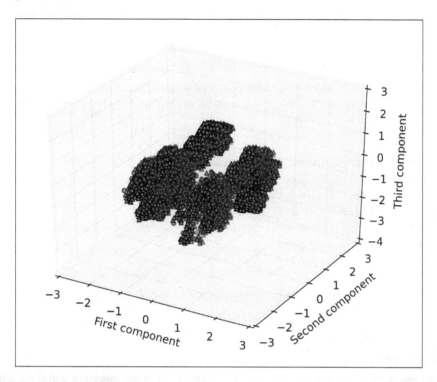

See also

Check this visualization of PCA at http://setosa.io/ev/principal-component-
analysis/.

Using Principal Component Analysis to find things that matter

Kernel PCA, in contrast to the PCA method that we just introduced, uses a user-defined kernel
function to map the dataset with n dimensions to an m-dimensional feature space. PCA uses
a linear function for the mapping and is equivalent to Kernel PCA with a linear kernel.

Kernel PCA can be especially useful if the data cannot be linearly separable so various
nonlinear kernels can be used to map your data to higher dimensions.

Getting ready

To execute this recipe, you will need `pandas` and `Scikit`. No other prerequisites are required.

How to do it...

Once again, we wrap our model in a method so that we can track how long it takes for the model to converge. With Kernel PCA, you should expect significantly longer estimation times (the `reduce_kernelPCA.py` file):

```
@hlp.timeit
def reduce_KernelPCA(x, **kwd_params):
    '''
        Reduce the dimensions using Principal Component
        Analysis with different kernels
    '''
    # create the PCA object
    pca = dc.KernelPCA(**kwd_params)

    # learn the principal components from all the features
    return pca.fit(x)
```

How it works...

As before, we first read in the dataset and split it into a set of independent variables x and the dependent variable y. Using the keyword `**kwd_params` parameter, we can test various kernel models with ease; in this example, we will use the RBF function introduced earlier in the book:

```
kwd_params = {
        'kernel': 'rbf',
        'eigen_solver': 'arpack',
        'n_components': 3,
        'max_iter': 1,
        'tol': 0.9,
        'gamma': 0.33
    }
```

The `kernel` parameter allows us to select different kernels: you can choose linear, poly, rbf, sigmoid, cosine, or precompute your own kernel. The n_components parameter controls how many principal components to find.

The `KernelPCA(...)` method iteratively finds the principal components in the data. The eigenvectors of the covariance matrix are found using the arpack solver.

To learn more about arpack, refer to `http://www.caam.rice.edu/software/ARPACK/` or `http://docs.scipy.org/doc/scipy/reference/tutorial/arpack.html`.

The `max_iter` parameter controls the maximum number of arpack iterations to run the estimation before stopping. This prevents situations when the estimation is stuck in a local extreme.

The `tol` parameter specifies to the tolerance level; this also controls the iterations. If the improvement between iterations is smaller than this value, the iterations will stop as the eigenvalues are found.

The gamma parameter is the coefficient for the `poly` and `rbf` kernels and is arguably the hardest parameter to choose wisely. To illustrate the influence of the gamma parameter on the RBF kernel, consider the following dataset (representing a XOR function):

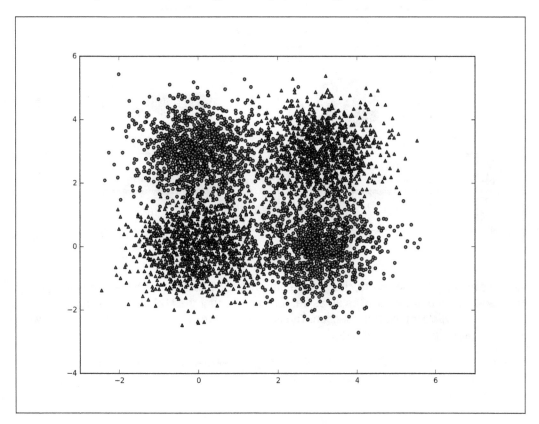

In this two-dimensional example, we want to find the best separation in the data. Here, we present the results of running Kernel PCA with an RBF kernel when we vary the value of the gamma parameter. Following the chart illustrating differing gamma parameters, we also present a plot of the first principal component only as a visual aid to check whether our data is separated or not:

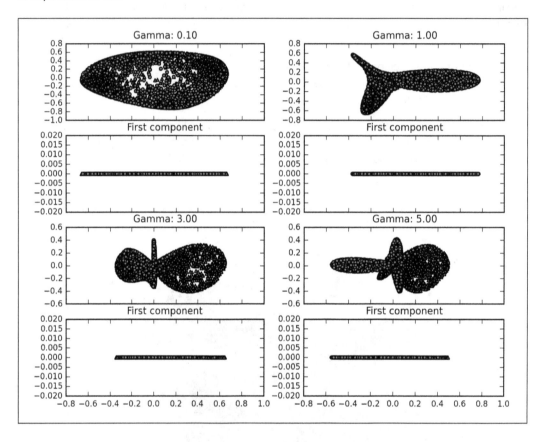

As you can see from the charts, the gamma parameter influences the shape and density of the transformed dataset. Also, as an aside, you can see that the XOR function is really hard to separate, even with the RBF function. The closest that we came to separating the data was with the gamma parameter set to 5.

The reduce_KernelPCA(...) method with the RBF kernel takes normally around 10 minutes on my machine to estimate.

There's more...

As mentioned earlier, we can use other functions as a kernel. In this recipe, we will use the XOR function dataset to present the idea how to estimate Kernel PCA with various kernel functions so that you do not have to wait as long as with the previous one (the `reduce_kernelPCA_alternative.py` file):

```
First, we prepare the dataset:
import sklearn.datasets as dt

# centers of the blobs
centers = [(0,0),(3,0),(3,3),(0,3)]

# create the sample
x, y = dt.make_blobs(n_samples = 5000, n_features=2,
    cluster_std=0.8, centers=centers, shuffle=False
)

# and make it XOR like
y[y == 2] = 0
y[y == 3] = 1
```

To start with, we import a helper module: `sklearn.datasets`. We want to create four blobs of data at four corners of a square, so we store the coordinates of these in the centers list. Using the `.make_blobs(...)` method, we create our dataset: we will have 5,000 data points (the `n_samples` parameter) with two features (as we want a two-dimensional dataset, `n_features`). The blobs of data will be fairly dense; this is controlled by the `cluster_std` parameter. We use the `centers` parameter to determine the central points of our blobs. The last parameter, shuffle, controls whether the dataset should be shuffled, hence introducing more variance.

The y object holds a list of labels for our data. As we passed four centers, we will get four distinct labels for our data, and our XOR dataset should have only two labels: the same for the blobs at diagonally opposite corners of the square. So, we change labels for two blobs.

Then, we prepare a list of all the kernels that we want to test:

```
kwd_params = [{ 'kernel': 'linear',
        'n_components': 2,'max_iter': 3,
        'tol': 1.0, 'eigen_solver': 'arpack'
    }, { 'kernel': 'poly',
        'degree': 2,'n_components': 2,'max_iter': 3,
        'tol': 1.0, 'eigen_solver': 'arpack'
    }, { 'kernel': 'sigmoid',
        'n_components': 2,'max_iter': 3,
        'tol': 1.0, 'eigen_solver': 'arpack'
```

```
    }, { 'kernel': 'cosine',
        'degree': 2,'n_components': 2,'max_iter': 3,
        'tol': 1.0, 'eigen_solver': 'arpack'}
    ]
```

Then, loop through all of them, estimating the models and saving charts:

```
color_marker = [('r','^'),('g','o')]

for params in kwd_params:
    z = reduce_KernelPCA(x, **params)

    # plot and save the chart
    # vary the colors and markers for the points
    file_save_params = {
        'filename':
        '../../Data/Chapter5/charts/kernel_pca_3d_{0}.png'\
            .format(params['kernel']),
        'dpi': 300
    }

    hlp.plot_components_2d(z.transform(x), y, color_marker,
        **file_save_params)
```

The linear kernel is the equivalent of running a normal PCA method. With the XOR dataset, I did not expect it to perform well as it is not linearly separable. The polynomial produces a somewhat better separation and so does the sigmoid. Cosine, however, still has some red dots within the green ones, although one can really discern a pattern. In case of cosine (as with all the other methods), you would not be able to separate the data with only one dimension. To see all the charts, look in the `Data/Chapter5/charts/` folder.

See also

Sebastian Raschka gives us some nice examples of Kernel PCA:

```
http://sebastianraschka.com/Articles/2014_kernel_pca.html
```

Finding the principal components in your data using randomized PCA

PCA (and Kernel PCA) both use low-rank matrix approximation to estimate the principal components. The low-rank matrix approximation minimizes a cost function represented as a fit between a given matrix and its approximation.

Such a method might be really costly for big datasets. By randomizing how the singular value decomposition of the input dataset happens, the speed up in the estimation is significant.

Getting ready

To execute this recipe, you will need `NumPy`, `Scikit`, and `Matplotlib`. No other prerequisites are required.

How to do it...

As before, we create a wrapper method to estimate our model (the `reduce_randomizedPCA.py` file):

```
def reduce_randomizedPCA(x):
    '''
        Reduce the dimensions using Randomized PCA algorithm
    '''
    # create the CCA object
    randomPCA = dc.RandomizedPCA(n_components=2, whiten=True,
        copy=False)

    # learn the principal components from all the features
    return randomPCA.fit(x)
```

How it works...

The inside of the method looks almost identical to all our previous model wrappers. First, we create the model object and return it upon fitting the data. `Scikit`, across its portfolio of decomposition methods, keeps a uniform **API (Application Programming Interface)**, so we do not need to change much to try other models.

In this recipe, we will focus on comparing how much speed up you can expect from randomized PCA versus PCA when your data grows in feature and sample size both. Hence, let's first define the parameters:

```
sampleSizes = np.arange(1000, 50000, 3000)
featureSpace = np.arange(100, 1000, 100)
```

The `.arange(<start>,<end>,<step>)` method of `NumPy` creates a range starting at `<start>` and finishing at (or around) `<end>` with a defined `<step>`.

In the `helper.py` file, we added a new method that is similar to our `.timeit` decorator:

```
def timeExecution(method, *args, **kwargs):
    '''
        A method to measure time of execution of a method
    '''
    start = time.time()
    result = method(*args, **kwargs)
    end = time.time()

    return result, end-start
```

The `timeExecution(...)` accepts method as its first parameter and whatever parameter this method accepts next (the `*args, **kwargs` set of variable size parameters). Inside, the method follows a similar pattern as our `timeit` decorator but returns not only the results from running the method, but also the time of execution.

We will use `timeExecution(...)` to measure the time of execution of PCA and randomized PCA, which we are going to plot later; `Z` is a dictionary that will hold the execution times:

```
Z = {'randomPCA': [], 'PCA': []}
```

The main loop of the script looks as follows:

```
for features in featureSpace:
    inner_z_randomPCA = []
    inner_z_PCA = []

    for sampleSize in sampleSizes:
        # get the sample
        x, y = hlp.produce_sample(
            sampleSize=sampleSize, features=features)

        print(
            'Processing: sample size {0} and {1} features'\
            .format(sampleSize, features))

        # reduce the dimensionality
        z_r, time_r    = hlp.timeExecution(
            reduce_randomizedPCA, x)
        z_pca, time_pca = hlp.timeExecution(
            reduce_PCA, x)

        inner_z_randomPCA.append(time_r)
        inner_z_PCA.append(time_pca)

    Z['randomPCA'].append(inner_z_randomPCA)
    Z['PCA'].append(inner_z_PCA)
```

We have two loops to go through as we compare the execution times along two dimensions. This is, ultimately, to answer what has a greater impact on the execution time: the number of features or sample size. The `inner_z_...` lists will hold the execution times.

As the first step in the second loop, we use the `produce_sample(...)` method. This is yet another method that we added to our `helper.py` file:

```
def produce_sample(sampleSize, features):
    import sklearn.datasets as dt

    # create the sample
    x, y = dt.make_sparse_uncorrelated(
        n_samples=sampleSize, n_features=features)

    return x, y
```

The method produces a sample with predefined `sampleSize` and number of features using the `make_sparse_uncorrelated` method of Scikit.

With the sample created, we can now time the execution of our `reduce_randomizedPCA(...)` and `reduce_PCA(...)` methods: we sourced the latter one from the *Reducing the dimensions using the kernel version of PCA* recipe.

There's more...

We then append the times to our inner lists that hold the execution times. The loop finishes by appending the z dictionary of lists.

Let's now plot the execution times:

```
def saveSurfacePlot(X_in, Y_in, Z, **f_params):
    from mpl_toolkits.mplot3d import Axes3D
    import matplotlib.pyplot as plt
    import matplotlib as mt

    # adjust the font
    font = {'size': 8}
    mt.rc('font', **font)

    # create a mesh
    X, Y = np.meshgrid(X_in, Y_in)

    # create figure and add axes
    fig = plt.figure()
    ax = fig.gca(projection='3d')
```

```
# plot the surface
surf = ax.plot_surface(X, Y, Z,
    rstride=1, cstride=1,
    cmap=mt.cm.seismic,
    linewidth=0, antialiased=True)

# set the limits on the z-axis
ax.set_zlim(0, 7)

# add labels to axes
ax.set_xlabel('Sample size')
ax.set_ylabel('Feature space')
ax.set_zlabel('Time to estimate (s)')

# rotate the chart
ax.view_init(30, 130)

# and save the figure
    fig.savefig(**f_params)

f_params = {
    'filename':'../../Data/Chapter5/charts/time_pca_surf.png',
    'dpi': 300
}

saveSurfacePlot(sampleSizes, featureSpace,
    Z['randomPCA'], **f_params)
```

The `saveSurfacePlot(...)` method first imports all the necessary modules that we will use. We reduce the size of the font so that the chart is more readable.

The `.meshgrid(...)` method of `NumPy` creates *n* x *m* matrices `X` and `Y`, where the size of `X_in` is m and size of `Y_in` is n; `X_in` repeats *n* times (rows) in `X` while `Y_in` repeats *m* times (columns) in `Y`. Effectively, when we iterate through all the *m* x *n* elements of each matrix, we cover all the possible combinations of features and sample sizes.

Next, we create a figure, add three-dimensional axes, and plot our surface. *X*, *Y*, and *Z* are the coordinates of the points. The `rstride` and `cstride` parameters specify that we want to plot the horizontal and vertical lines along the surface respectively, originating from each point on the *x* and *y* axes; doing so creates tiles on the surface of the plane. The `linewidth` parameter controls the width of the line while the `cmap` parameter defines the color map (or temperature map)—the higher the value of `z`, the redder the plot will get.

 You can find a list of all the available color maps here:
`http://matplotlib.org/examples/color/colormaps_reference.html`

The `antialiased` parameter creates smoother lines on the chart.

We set the limit on the *z* axis so that when we compare the charts for the two different methods, we can immediately spot the differences in execution times. Adding labels always aids in understanding what our chart presents.

 You might have to adjust the `.set_zlim(...)` parameters as the execution times on your machine will most likely differ.

Lastly, we rotate the chart and save the figure. The `.view_init(...)` method sets the elevation at 30 degrees and rotates the chart by 130 degrees clockwise in the *x-y* plane:

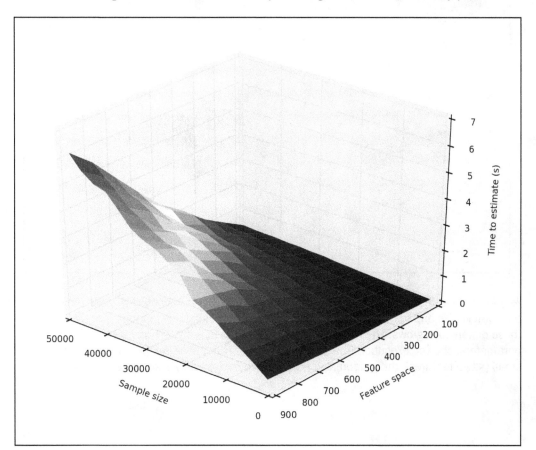

The plot for PCA should look similar to the preceding one, while for the randomized version, it should look more like the following one:

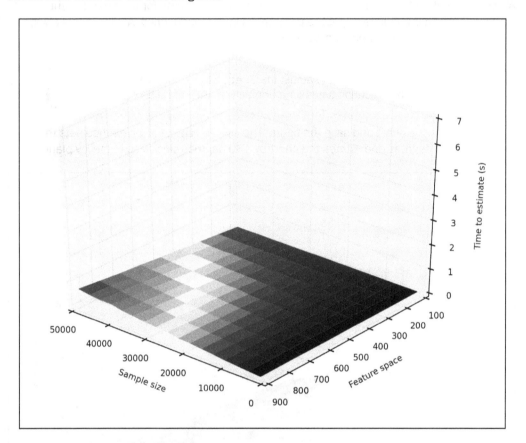

As you can see, the execution time of the PCA algorithm grows very quickly when both the sample size and feature space grow; it, however, remains fairly immune if only one of these dimensions grows. The randomized version of the PCA algorithm, for the same set of parameters, showed execution times 10 times faster. Note, however, that if your dataset is small (<100 features), the randomized PCA method can actually take longer.

Extracting the useful dimensions using Linear Discriminant Analysis

Now that we understand the mechanics (and trade-offs) of dimensionality reduction, let's use it for classification.

In this recipe, we will introduce **Linear Discriminant Analysis (LDA)**. LDA, in contrast to the methods presented earlier in this chapter, aims at representing the dependent variable as a linear function of many other features; in that sense, it is similar to a regression (which we will discuss in the next chapter). The LDA shows similarities to the ANOVA analysis of variance and logistic regression in how it models the linear relationships in the data that capture (explain) the variance the best.

We will use a linear SVM classifier to test the effectiveness of our dimensionality reduction efforts compared to fitting the classifier with the original dataset. We will, once again, use the marketing calls of a bank dataset.

Getting ready

To run this recipe, you will need `pandas` and `MLPY`. No other prerequisites are required.

How to do it...

To reduce the dimensions using LDA, you can use the following method (the `reduce_LDA.py` file):

```
@hlp.timeit
def reduce_LDA(x, y):
    '''
        Reduce the dimensions using Linear Discriminant
        Analysis
    '''
    # create the PCA object
    lda = ml.LDA(method='fast')

    # learn the principal components from all the features
    lda.learn(x, y)

    return lda
```

How it works...

The execution, as before, starts by reading in the dataset and splitting it into a set of independent variables x and the dependent variable y. As, in this recipe, we want to compare the impact of reducing the dimensions on our classification efforts, we also split the original dataset into training and testing subsets:

```
train_x_orig, train_y_orig, \
test_x_orig,  test_y_orig, \
labels_orig = hlp.split_data(
    csv_read,
    y = 'credit_application'
)
```

Having done so, now we can reduce the dimensionality of our dataset:

```
csv_read['reduced'] = reduce_LDA(x, y).transform(x)
```

There are a couple of things happening here in one go: let's analyze these one by one.

First and foremost, we call the reduce_LDA(...) method. The method requires two parameters to be passed, x, which is our set of independent variables, and y, our dependent variable. This is somewhat unorthodox, as all the previous methods required only the independent variables. However, as LDA is trying to find (model) linear relationships between the set of independent variables and the dependent one, it requires to know the *target* (you should remember the *training with a teacher paradigm* we talked about at the beginning of *Chapter 4, Clustering Techniques* in the introduction).

MLPY provides the .LDA(...) estimator. Here, we specify the method to be fast as our dataset has over 40 k records. Next, we ask the model to .learn(...) the relationships in our data and return the trained object.

The .transform(x) method encodes the set of independent variables, which we then store in the reduced column in our original dataset, csv_read.

Now, it is time to create another training and testing dataset for our classifier:

```
train_x_r, train_y_r, \
test_x_r,  test_y_r, \
labels_r = hlp.split_data(
    csv_read,
    y = 'credit_application',
    x = ['reduced']
)
```

Here, we only select the reduced column as our independent variable.

We now proceed to estimating our classifiers:

```
classifier_r    = fitLinearSVM((train_x_r, train_y_r))
classifier_orig = fitLinearSVM((train_x_orig, train_y_orig))
```

No surprises here, as you should remember how we have done this throughout *Chapter 3, Classification Techniques*. Here, we reuse the `fitLinearSVM(...)` method that we introduced in *Chapter 3, Classification Techniques* in the *Utilizing Support Vector Machines as a classification engine* recipe:

```
@hlp.timeit
def fitLinearSVM(data):
    '''
        Build the linear SVM classifier
    '''
    # create the classifier object
    svm = ml.LibSvm(svm_type='c_svc',
        kernel_type='linear', C=100.0)

    # fit the data
    svm.learn(data[0],data[1])

    # return the classifier
    return svm
```

Let's test how well our methods did:

```
# classify the unseen data
predicted_r    = classifier_r.pred(test_x_r)
predicted_orig = classifier_orig.pred(test_x_orig)

# print out the results
hlp.printModelSummary(test_y_r, predicted_r)
hlp.printModelSummary(test_y_orig, predicted_orig)
```

First, we predict the classes using the just estimated classifiers and then assess the models' efficacy. Here's what we saw:

```
The method reduce_LDA took 0.09 sec to run.
The method fitLinearSVM took 2.05 sec to run.
The method fitLinearSVM took 76.46 sec to run.
Overall accuracy of the model is 90.78 percent
Classification report:
               precision    recall  f1-score   support

         0.0       0.92      0.98      0.95     11957
         1.0       0.69      0.35      0.46      1538

avg / total        0.89      0.91      0.89     13495

Confusion matrix:
 [[11714    243]
 [ 1001    537]]
ROC:   0.664415961487
Overall accuracy of the model is 90.57 percent
Classification report:
               precision    recall  f1-score   support

         0.0       0.92      0.98      0.95     11930
         1.0       0.64      0.34      0.45      1492

avg / total        0.89      0.91      0.89     13422

Confusion matrix:
 [[11645    285]
 [  981    511]]
ROC:   0.659301971509
```

The LDA was really fast on our dataset.

You can notice, however, the difference in execution times: the model estimated on the full (not reduced) dataset took close to 50 times longer than the one using reduced dimensions. This should not really surprise you as the reduced dataset had only one feature while the full one had 59.

However, when you compare the power of both models, the one with the reduced feature space outperformed the one with the full dataset! The difference is not enormous, but the SVM estimated with the reduced features performed better in terms of recall, f-score, and ROC. Also, the overall accuracy was closer to 91% while for the dataset with the full feature set was closer to 90%.

Using various dimension reduction techniques to classify calls using the k-Nearest Neighbors classification model

Now that we have seen that reducing dimensions can lead to better performing classification models, let's try a couple of more methods and introduce another classification algorithm: the **k-Nearest Neighbors (kNN)** algorithm.

In this recipe, we will test and compare three dimensionality reduction methods: PCA (as a benchmark), fast **Independent Component Analysis (ICA)**, and the truncated SVD method.

Getting ready

To execute this recipe, you will need `pandas` and `Scikit`. No other prerequisites are required.

How to do it...

In this recipe, we leverage the fact that everything in Python is an object (methods as well) and we can pass these around to other methods as parameters. Here is how we will reduce dimensions in this recipe (the `reduce_kNN.py` file):

```
@hlp.timeit
def reduceDimensions(method, data, **kwrd_params):
    '''
        Reduce the dimensions
    '''
    # split into independent and dependent features
    x = data[data.columns[:-1]]
    y = data[data.columns[-1]]

    # create the reducer object
    reducer = method(**kwrd_params)

    # fit the data
    reducer.fit(x)

    # return the classifier
    return reducer.transform(x)
```

How it works...

The `reduceDimensions(...)` method accepts the `reducer` method as its first parameter, data to be used in the estimation comes second, and all the keyword arguments for the reducer method are passed as the third argument.

We moved the data splitting code in our method to reduce the code redundancy. As with our previous model-specific feature reduction methods, we first create reducer and then fit the data. This time, however, we return already transformed data.

For the purpose of this recipe, we developed a set of model-specific methods that handle the data reduction and classification for different reducers. Let's take the method that reduces the feature space, using PCA as an example:

```
def fit_pca(data):
    kwrd_params = {'n_components': 5, 'whiten': True}

    reduced = reduceDimensions(cd.PCA, data, **kwrd_params)

    data_l = prepare_data(data, reduced,
        kwrd_params['n_components'])

    class_fit_predict_print(data_l)
```

This method only requires that we pass the data. First, we define the parameters specific to the method that we want to use. Next, we call our `reduceDimensions(...)` method passing a specific reducer method; in this case, `cd.PCA` and all the other required parameters, `data` and `**kwrd_params`.

We then prepare the dataset required to estimate the kNN classifier. To this end, we use the `prepare_data(...)` method:

```
def prepare_data(data, principal_components, n_components):
    cols = ['pc' + str(i)
        for i in range(0, n_components)]

    data = pd.concat(
        [data,
            pd.DataFrame(principal_components,
                columns=cols)],
            axis=1, join_axes=[data.index])

    # split the data into training and testing
    train_x, train_y, \
    test_x,  test_y, \
    labels = hlp.split_data(
        data,
```

```
        y = 'credit_application',
        x = cols
    )


    return (train_x, train_y, test_x, test_y)
```

The method accepts data as its first argument, an encoded list (matrix) of principal components, and number of components that we expect to have. In all of our modeling efforts in this recipe, we instruct the dimensionality reduction models to return five principal components.

First, we create a list of column names so that we can then concatenate our original dataset with the reduced matrix. We do this using the `.concat(...)` method of `pandas`. The method accepts a list of DataFrames to be concatenated as the first parameter; here, we pass our original data and create a DataFrame from `principal_components` with specified columns. The axis parameter to `.concat(...)` instructs the join direction (rows) and `join_axes` specifies the join key.

Now that we have extended the original dataset with the principal components' columns, we can split the dataset into training and testing. Finally, we return a tuple of training and testing subsets.

With the data prepared, we can now estimate the kNN and evaluate the model by calling the `class_fit_predict_print(...)` method in the `fit_pca(...)` method:

```
@hlp.timeit
def fit_kNN_classifier(data):
    '''
        Build the kNN classifier
    '''
    # create the classifier object
    knn = nb.KNeighborsClassifier()

    # fit the data
    knn.fit(data[0],data[1])

    #return the classifier
    return knn

def class_fit_predict_print(data):
    # train the model
    classifier = fit_kNN_classifier((data[0], data[1]))

    # classify the unseen data
    predicted = classifier.predict(data[2])

    # print out the results
    hlp.printModelSummary(data[3], predicted)
```

We pass in the training and testing data (in the form of a (`train_x, train_y, test_x, test_y`) tuple).

First, we use the `fit_kNN_classifier(...)` method to estimate our classifier. We only pass a tuple of `(train_x, train_y)` as the datasets to be used in the estimation.

As with other estimation methods, we first create the classifier object (`.KNeighborsClassifier(...)`).

> There are plenty of parameters that you can use in `.KNeighborsClassifier(...)`, but we decided to keep it simple. You can check the method's documentation at `http://scikit-learn.org/stable/modules/generated/sklearn.neighbors.KNeighborsClassifier.html`.

Then, we fit the model and return the classifier. We use the `.predict(...)` method that `.KNeighborsClassifier(...)` exposes to predict the classes for our `test_x` dataset. Finally, we produce the model's summary using the `.printModelSummary(...)` method, which we developed in the *Testing and comparing the models* recipe in *Chapter 3, Classification Techniques*, with the true ones (`test_y`).

As mentioned at the beginning, in this recipe, we compare the efficacy of the kNN model when estimated using an unreduced dataset with data having a reduced feature space:

```
# compare models
fit_clean(csv_read)
fit_pca(csv_read)
fit_fastICA(csv_read)
fit_truncatedSVD(csv_read)
```

Here are the results:

▸ `fit clean(...)`

```
The method fit_kNN_classifier took 0.63 sec to run.
Overall accuracy of the model is 89.18 percent
Classification report:
                precision    recall  f1-score   support

         0.0        0.91      0.98      0.94     12075
         1.0        0.55      0.24      0.33      1539

avg / total        0.87      0.89      0.87     13614

Confusion matrix:
 [[11777   298]
 [ 1175   364]]
ROC:   0.605919064973
```

▶ `fit_pca(...)`

```
The method reduceDimensions took 0.14 sec to run.
The method fit_kNN_classifier took 0.02 sec to run.
Overall accuracy of the model is 91.82 percent
Classification report:
               precision    recall   f1-score    support

         0.0       0.93      0.98       0.95      12171
         1.0       0.76      0.44       0.56       1610

 avg / total       0.91      0.92       0.91      13781

Confusion matrix:
 [[11948    223]
 [  904    706]]
ROC:  0.710093537688
```

▶ `fit_fastICA(...)`

```
The method reduceDimensions took 0.21 sec to run.
The method fit_kNN_classifier took 0.02 sec to run.
Overall accuracy of the model is 91.74 percent
Classification report:
               precision    recall   f1-score    support

         0.0       0.93      0.98       0.95      12112
         1.0       0.72      0.43       0.54       1535

 avg / total       0.91      0.92       0.91      13647

Confusion matrix:
 [[11858    254]
 [  873    662]]
ROC:  0.705149710197
```

▶ `fit_truncatedSVD(...)`

```
The method reduceDimensions took 0.09 sec to run.
The method fit_kNN_classifier took 0.02 sec to run.
Overall accuracy of the model is 93.15 percent
Classification report:
              precision    recall  f1-score   support

         0.0       0.94      0.98      0.96     12063
         1.0       0.78      0.53      0.63      1499

avg / total         0.93      0.93      0.93     13562

Confusion matrix:
[[11846    217]
 [  712    787]]
ROC:  0.753513893067
```

The `fit_clean(...)` method fits a kNN on a fully featured dataset. In comparison, it performed the worst out of all the methods tested, attaining only 89% overall accuracy; the other methods were much closer to the 92% mark, with the truncated SVD attaining almost 93%. The latter one has also outperformed all the other methods in terms of the precision, recall, and f-score as well as ROC.

6
Regression Methods

In this chapter, we will cover various techniques to predict the output of a power plant. You will learn the following recipes:

- ▶ Identifying and tackling multicollinearity in your data
- ▶ Building a linear regression model to predict the power plant output
- ▶ Using OLS to forecast how much electricity can be produced
- ▶ Estimating the output of an electric plant using CART
- ▶ Employing the kNN model in a regression problem
- ▶ Applying the Random Forest model to a regression analysis
- ▶ Gauging the amount of electricity a plant can produce using SVMs
- ▶ Training a Neural Network to predict the output of a power plant

Introduction

One of the most common problems in the real world is to predict certain quantities or, in more general terms, find a relationship between a set of independent variables and the dependent one. In this chapter, we will focus on predicting the output of a power plant.

The dataset that we will use in this chapter comes from the U.S. Energy Information Administration. We procured the 2014 data from their website, `http://www.eia.gov/electricity/data/eia923/xls/f923_2014.zip`.

We will use the data from the `EIA923_Schedules_2_3_4_5_M_12_2014_Final_Revision.xlsx` file only, sheet `Generation and Fuel Data`. We will be predicting Net Generation (Megawatt hours). As most of the data is categorical (state or fuel type), we decided to dummy code them.

Ultimately, our dataset holds only a subset of 4,494 records of the whole dataset. We selected only the power plants with an output greater than 100 MWh in 2014 that were located in a handful of selected states. We also only selected plants that use certain types of fuel—Aggregate Fuel Code (AER): coal (COL), distillate (DFO), hydroelectric conventional (HYC), biogenic municipal solid waste and landfill gas (MLG), natural gas (NG), nuclear (NUC), solar (SUN), and wind (WND). In addition, only plants with certain movers were selected: combined-cycle combustion turbine part (CT), combustion turbine (GT), hydraulic turbine (HY), internal combustion (IC), photovoltaic (PV), steam turbine (ST), and wind turbine (WT).

The data output of power plants varies greatly and is heavily skewed, with median output of 18,219 MWh and average of 448,213 MWh; 25% of all the power plants (2,017 plants) produced less than 3,496 MWh in 2014, with an average per capita use of electric energy in the United States of around 12 kWh (2010 data, http://energyalmanac.ca.gov/electricity/us_per_capita_electricity-2010.html), which equates to a town of 291 people, which is depicted in the following screenshot:

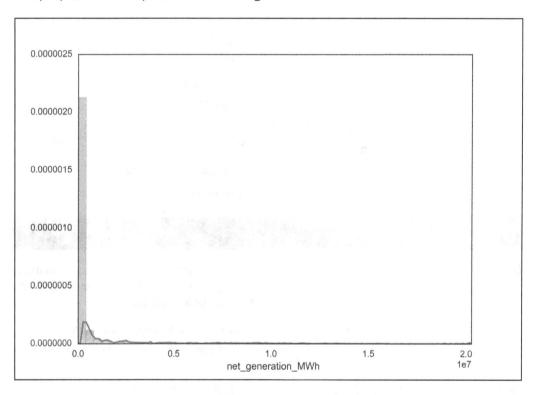

Most of the power plants use natural gas (over **1,300**), approximately **750** were hydroelectric or photovoltaic, and around **580** were powered by wind. Thus, a fair proportion of power plants in our dataset were powered from renewable energy sources:

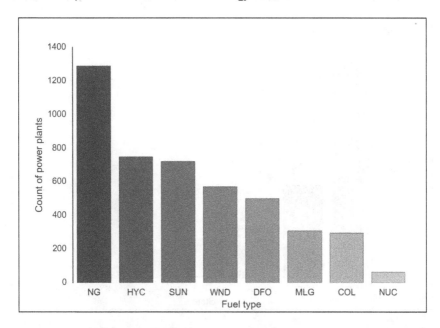

Close to **2,500** power plants used steam turbine to generate electricity, followed by hydraulic turbine and combustion turbine:

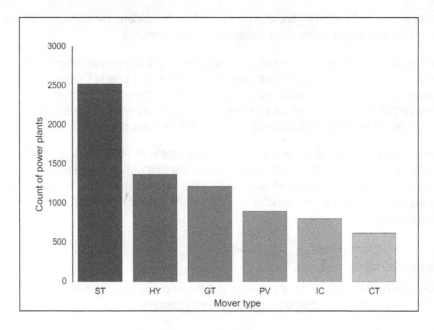

California is the state with the highest number of power plants in our dataset, with over **1,200** of them. Texas comes far second with only over **500**, followed by Massachusetts:

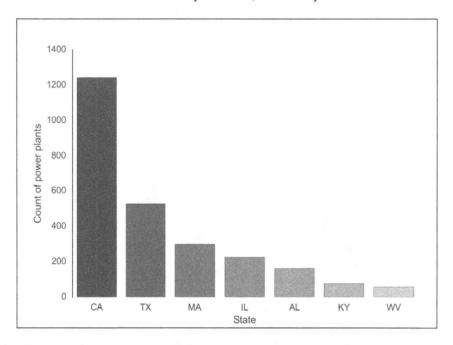

Identifying and tackling multicollinearity

Multicollinearity is a situation where one (or more) of independent variables can be expressed as a linear combination of some other independent variables.

For example, consider a situation where we try to predict the power consumption for a state using population, number of households, and number of power plants located in the state. In a situation like this, one might clearly deduce that the more people living in the state, the higher number of households one might expect, that is, the number of households can be represented by some (close to) linear relationship of the state's population.

Now, if we were to estimate a model based on a data that is collinear, very good chances are that one (or even all the variables that are collinear) will turn out as insignificant. In contrast, removing the collinear variables (and keeping only the variable that is the most correlated with our dependent variable, that is, explains most of its variation) will not reduce the model's explanatory power.

Getting ready

To execute this recipe, you will need `pandas`, `NumPy`, and `Scikit`; we will use Scikit to reduce the dimensions. No other prerequisites are necessary.

How to do it...

To find out if our data is multicollinear, we need to inspect the eigenvalues of the correlation matrix of independent variables; if the data is collinear, analyzing the correlation matrix will help us identify the variables (the `regression_multicollinearity.py` file):

```
import sys
sys.path.append('..')

import helper as hlp
import pandas as pd
import numpy as np
import sklearn.decomposition as dc

# the file name of the dataset
r_filename = '../../Data/Chapter6/power_plant_dataset.csv'

# read the data
csv_read = pd.read_csv(r_filename)

x = csv_read[csv_read.columns[:-2]].copy()
y = csv_read[csv_read.columns[-1]]

# produce correlation matrix for the independent variables
corr = x.corr()

# and check the eigenvectors and eigenvalues of the matrix
w, v = np.linalg.eig(corr)
print('Eigenvalues: ', w)

# values that are close to 0 indicate multicollinearity
s = np.nonzero(w < 0.01)
# inspect which variables are collinear
print('Indices of eigenvalues close to 0', s[0])

all_columns = []
for i in s[0]:
    print('\nIndex: {0}. '.format(i))

    t = np.nonzero(abs(v[:,i]) > 0.33)
    all_columns += list(t[0])
    print('Collinear: ', t[0])
```

How it works...

First, we load all the necessary modules and read in the dataset in the usual fashion. Next, we split the dataset into a set of independent variables x (we create a copy of the original dataset) and a dependent variable y.

Next, we produce a matrix of correlations between all the independent variables and find the eigenvalues and eigenvectors using NumPy's eig(...) method from its linalg module. The eig(...) method requires a square matrix as its input.

 Eigenvectors with corresponding eigenvalues can be found only for square matrices. For non-square matrices, singular values can be found (https://www.math.washington.edu/~greenbau/Math_554/Course_Notes/ch1.5.pdf).

The eigenvalues that are very close to 0 indicate (multi)collinearity. To find the indices of the values that are close to 0 (in our case, <0.01), we first create a truth vector: a vector where each element is either False (0) or True (1). This can be accomplished by simply stating w < 0.001; the result of such an operation is the truth vector of equal length to the original vector w, where each element of the original vector that is smaller than 0.001 is flagged as True. Now we can use the .nonzero(...) method of NumPy that returns a list of indices of all non-zero elements (in our case, the locations of the elements equal to True). There is a good chance that you might get eigenvalues close to 0 at more than one location. For our dataset, you can expect something similar to the following:

```
Indices of eigenvalues close to 0: [28 29 30 31 32]
```

We loop through all the indices and find the corresponding eigenvectors. Elements of the eigenvectors that are significantly greater than 0 help us find the variables that are collinear. As the elements of the eigenvector can have negative values, we return all the indices that have the absolute value > 0.33:

```
Index: 28.
Collinear:   [0 1 3]
Index: 29.
Collinear:   [ 9 11 12 13]
Index: 30.
Collinear:   [15]
Index: 31.
Collinear:   [ 2  4 10 14]
Index: 32.
Collinear:   [ 4 14]
```

Note that these variables repeat. We store all these values in the `all_columns` list. Once we have gone through all the eigenvalues that were close to `0`, let's check which variables are collinear:

```
for i in np.unique(all_columns):
    print('Variable {0}: {1}'.format(i, x.columns[i]))
```

We loop through all distinct values stored in `all_columns` (using the `.unique(...)` method of `NumPy`) and print out the corresponding variable name from the list of all columns of `x`:

```
Variable 0: fuel_aer_NG
Variable 1: fuel_aer_DFO
Variable 2: fuel_aer_HYC
Variable 3: fuel_aer_SUN
Variable 4: fuel_aer_WND
Variable 9: mover_GT
Variable 10: mover_HY
Variable 11: mover_IC
Variable 12: mover_PV
Variable 13: mover_ST
Variable 14: mover_WT
Variable 15: state_CA
Variable 28: state_OH
Variable 29: state_GA
Variable 30: state_WA
Variable 31: total_fuel_cons
Variable 32: total_fuel_cons_mmbtu
```

It should be clear now that we see collinearity in our dataset: certain types of movers will definitely be used with certain types of fuels. As an example, the wind turbine will more than likely use wind to move its propeller, photovoltaic mover will use sun, and hydraulic turbine will be propelled by water. Also, the state of California shows up; California is the state with the largest number of wind turbines (followed by Texas, `http://www.awea.org/resources/statefactsheets.aspx?itemnumber=890`).

There's more...

One way to tackle the multicollinearity (apart from removing the variables from the model) is to reduce the dimensionality of the dataset. We can use PCA to do this for us (refer to the *Reducing the dimensions using the kernel version of PCA* recipe from *Chapter 5, Reducing Dimensions*).

We will reuse the `reduce_PCA(...)` method:

```
n_components = 5
z = reduce_PCA(x, n=n_components)
```

```
pc = z.transform(x)

# how much variance each component explains?
print('\nVariance explained by each principal component: ',
    z.explained_variance_ratio_)

# and total variance accounted for
print('Total variance explained: ',
    np.sum(z.explained_variance_ratio_))
```

We want to get the first five **principal components** (**PCs**). The `reduce_PCA(...)` method estimates the PCA model and the `.transform(...)` method transforms our data and returns the five principal components. We can check how much of the overall variance is explained by each principal component:

```
Variance explained by each principal component:   [ 0.1448439
  0.13074582   0.11066929   0.09373292   0.0670638 ]
```

So the first principal component explains around 14.5% of the variance, the second one 13%, and the third one around 11%. The two remaining PCs contribute together around 16% so the total variance explained is as follows:

```
Total variance explained:   0.547055726325
```

We store the principal components along with all the other variables in a file:

```
# append the reduced dimensions to the dataset
for i in range(0, n_components):
    col_name = 'p_{0}'.format(i)
    x[col_name] = pd.Series(pc[:, i])

x[csv_read.columns[-1]] = y
csv_read = x

# output to file
w_filename = '../../Data/Chapter6/power_plant_dataset_pc.csv'
with open(w_filename, 'w') as output:
    output.write(csv_read.to_csv(index=False))
```

First, we append the PCs to the back of our x dataset; we want to keep our dependent variable as last. Then, we output the dataset to `power_plant_dataset_pc.csv` that we will use in the following recipes.

Building Linear Regression model

Linear regression is arguably the simplest model that one can build. It is a model of choice if you know that the relationship between your dependent variable and independent ones is linear.

Getting ready

To execute this recipe, you will need `pandas`, `NumPy`, and `Scikit`. No other prerequisites are required.

How to do it...

Estimating the regression model with `Scikit` is extremely easy (the `regression_linear.py` file):

```python
import sys
sys.path.append('..')

# the rest of the imports
import helper as hlp
import pandas as pd
import numpy as np
import sklearn.linear_model as lm

@hlp.timeit
def regression_linear(x,y):
    '''
        Estimate a linear regression
    '''
    # create the regressor object
    linear = lm.LinearRegression(fit_intercept=True,
        normalize=True, copy_X=True, n_jobs=-1)

    # estimate the model
    linear.fit(x,y)

    # return the object
    return linear

# the file name of the dataset
r_filename = '../../Data/Chapter6/power_plant_dataset_pc.csv'
```

```python
# read the data
csv_read = pd.read_csv(r_filename)

# select the names of columns
dependent = csv_read.columns[-1]
independent_reduced = [
    col
    for col
    in csv_read.columns
    if col.startswith('p')
]

independent = [
    col
    for col
    in csv_read.columns
    if      col not in independent_reduced
        and col not in dependent
]

# split into independent and dependent features
x     = csv_read[independent]
x_red = csv_read[independent_reduced]
y     = csv_read[dependent]

# estimate the model using all variables (without PC)
regressor = regression_linear(x,y)

# print model summary
print('\nR^2: {0}'.format(regressor.score(x,y)))
coeff = [(nm, coeff)
    for nm, coeff
    in zip(x.columns, regressor.coef_)]
intercept = regressor.intercept_
print('Coefficients: ', coeff)
print('Intercept', intercept)
print('Total number of variables: ',
    len(coeff) + 1)
```

How it works...

As always, we first import all the necessary modules and read in our dataset.

We then split our data into subsets. First, we select the names of the columns that indicate the dependent variable (which is the last column in our dataset), principal component variables (`independent_reduced`), and the remaining independent variables. To select the principal components, as all the principal component columns in our dataset start with `p_` (see the *Identifying and tackling multicollinearity* recipe, in the *There's more...* section), we use Python's `.startswith(...)` built-in string method. To select the independent variables, we select all the columns that are not in either `independent_reduced` or `dependent`. We then extract these certain columns from the main dataset.

To estimate the model, we call the `regression_linear(...)` method. The method accepts two parameters: `x`, which is a set of our independent variables, and `y`, which is our dependent variable.

In the `regression_linear(...)` method, we first use the `.LinearRegression(...)` method to create the model object. We specify that the model should estimate the constant (`fit_intercept=True`), normalize the data, and copy the independent variables.

 There is always a possibility that a method might change the data passed to it as Python (by default) passes the parameters to methods by reference (not by value, like C or C++) http://www.python-course.eu/passing_arguments.php.

We also specify the number of jobs to spin to be equal to the number of cores that your computer has. For linear regression, this should rarely be a problem, though.

Once we have the model object, we `.fit(...)` the data, that is, estimate the model, and return the estimated model.

The performance of a linear regression model can be expressed in terms of R^2. The metric can be understood as a measure of explained variance by the model.

 In the recipe *Using OLS to forecast how much electricity can be produced*, we will present how to calculate R^2 by hand.

The `LinearRegression` object provides the `.score(...)` method. For our full model, we get the following:

```
R^2: 0.9965754630293856
```

To print the list of all the variables with their corresponding coefficients, we create the `coeff` list using list comprehension. We first `zip(...)` the columns of `x` with the `coeff_` attribute of our regressor; this creates a list of tuples that we subsequently iterate through to create our `coeff` list:

```
Coefficients:  [('fuel_aer_NG', -1042843895.0406576), ('fuel_aer_DFO',
-1042843895.040074), ('fuel_aer_HYC', 175530629.93238467), ('fuel_
aer_SUN', -1042843895.0388285), ('fuel_aer_WND', 11794696965.844255),
('fuel_aer_COL', -1042843895.0458133), ('fuel_aer_MLG',
-1042843895.0413487), ('fuel_aer_NUC', -1042843895.0566163), ('mover_
CT', 988883075.12895977), ('mover_GT', 988883075.12905121), ('mover_
HY', -229491449.84322524), ('mover_IC', 988883075.12988734), ('mover_
PV', 988883075.12761605), ('mover_ST', 988883075.12794578), ('mover_
WT', -11848657785.75465), ('state_CA', 2636968464.7743926), ('state_
TX', 2636968464.7731013), ('state_NY', 2636968464.7743831), ('state_
FL', 2636968464.7742429), ('state_MN', 2636968464.7741256), ('state_
MI', 2636968464.7749491), ('state_NC', 2636968464.7749896), ('state_
PA', 2636968464.7745223), ('state_MA', 2636968464.7743583), ('state_
WI', 2636968464.7746553), ('state_NJ', 2636968464.7743707), ('state_
IA', 2636968464.7738571), ('state_IL', 2636968464.7730536), ('state_
OH', 2636968464.774816), ('state_GA', 2636968464.7749362), ('state_
WA', 2636968464.7753024), ('total_fuel_cons', -0.067937591675115677),
('total_fuel_cons_mmbtu', 0.98815356874740379)]
```

Notice how big (and extremely minimal) the differences between coefficients for the states are. Also, all the power-related variables (fuel and mover) are negative. This, in itself, shows that state has a minimal impact on the overall performance of our model, especially when compared with the value of the intercept:

```
Intercept -2583007644.86
```

Also, we already know from the previous recipe that total fuel consumption would be correlated so we can safely remove one of the columns from the dataset.

Remembering that our dataset is poised with multicollinearity, let's estimate a model with PCs only:

```
regressor_red = regression_linear(x_red,y)

# print model summary
print('\nR^2: {0}'.format(regressor_red.score(x_red,y)))
coeff = [(nm, coeff)
    for nm, coeff
    in zip(x_red.columns, regressor_red.coef_)]
intercept = regressor_red.intercept_
print('Coefficients: ', coeff)
print('Intercept', intercept)
print('Total number of variables: ',
    len(coeff) + 1)
```

The model attains worse results in terms of R^2:

```
R^2: 0.12001001550007238
```

As the state variables had little impact, we decided to test the model without them. We also ditched `total_fuel_consumption` from the dataset:

```
columns = [col for col in independent if 'state' not in col and col !=
'total_fuel_cons']
x_no_state = x[columns]
```

The R^2 fell but insignificantly:

```
R^2: 0.9957289253419234
```

However, now we have a model with only 18 independent variables instead of 35 (including the intercept). As a rule of thumb, you should always opt for models that attain similar performance with a smaller number of explanatory variables.

There's more...

If you want to find the best line that fits your data (and you only have one explanatory variable), you can find the line that fits best using `Seaborn`. We will use our principal components to illustrate this although the results are not very informative.

First, as we store the principal components in columns, we will stack them one on top of the other so that we have only three columns in our dataset—`PC` to denote which principal component we plot, `x` for the value of the principal component, and `y` for the normalized value of the power plant output:

```
# stack up the principal components
pc_stack = pd.DataFrame()

# stack up the principal components
for col in x_red.columns:
    series = pd.DataFrame()
    series['x'] = x_red[col]
    series['y'] = y
    series['PC'] = col
    pc_stack = pc_stack.append(series)
```

First, we create an empty `pandas` DataFrame. Then, we loop through all the columns in `x_red` (that holds principal components only). The inner series DataFrame holds the values of the principal component as `x`, value of the dependent variable as `y`, and the name of the principal component as `PC`. We then append the series DataFrame to `pc_stack`.

Now we can plot the data:

```
sns.lmplot(x='x', y='y', col='PC', hue='PC', data=pc_stack,
           col_wrap=2, size=5)

pl.savefig('../../Data/Chapter6/charts/regression_linear.png',
    dpi=300)
```

The `.lmplot(...)` method fits the linear regression to our data. The `col` parameter specifies the grouping of our charts, that is, we will produce charts for our five principal components. The `hue` parameter will alter the colors of the charts for each principal component. The `col_wrap=2` will only place two charts in a row, each of which will have a size of five inches (height).

The resulting chart looks as follows:

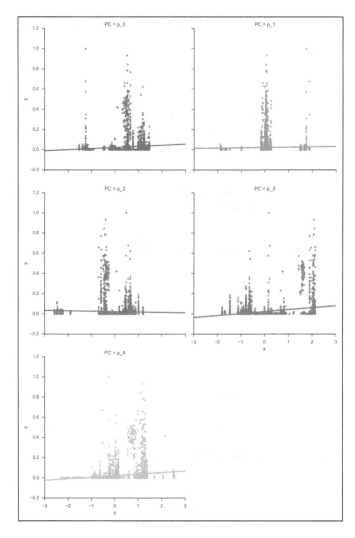

We also looked at the two total fuel variables to confirm if any linear relationship exists:

```
fuel_cons = ['total_fuel_cons','total_fuel_cons_mmbtu']
x = csv_read[fuel_cons]
```

The revealed chart confirms a linear relationship (as expected) between `total_fuel_cons_mmbtu` and an almost linear relationship for `total_fuel_cons`:

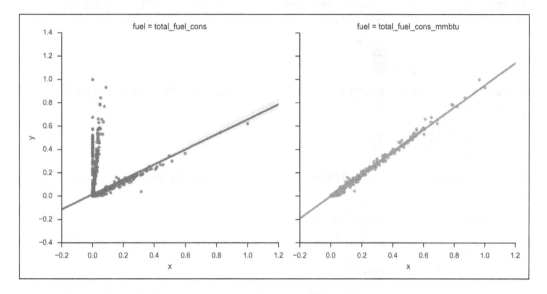

The total fuel consumption is the total amount of fuel consumed, irrespective of how much energy is stored in a unit of mass; it only tracks the total volume (or weight) of the resource used. However, one pound of uranium or plutonium will have much greater specific energy compared to one pound of coal. Thus, `total_fuel_cons` deviates strongly for lower values of x. When we take into account the specific energy (or energy density for liquid fuels), `total_fuel_cons_mmbtu`, we then see an almost perfect linear relationship between the energy consumption and energy generation.

As you can see from the chart, the trend of this line is < 1 indicating losses in the system (as expected). Note, however, that some of the fuel used is renewable, that is, we bear almost no input-related (fuel) operating cost to run wind turbines or solar panels.

Using OLS to forecast how much electricity can be produced

Ordinary Least Squares (**OLS**) is also a linear model. In fact, a linear regression is estimated using the least squares method as well. The OLS is, however, capable of estimating a model where the relationship between the dependent and independent variables is nonlinear as long as this relationship is linear in parameters.

Getting ready

To execute this recipe, you will need `pandas` and `Statsmodels`. No other prerequisites are required.

How to do it...

We will, as always, wrap our model estimation efforts in a function (the `regression_ols.py` file):

```python
import statsmodels.api as sm

@hlp.timeit
def regression_ols(x,y):
    '''
        Estimate a linear regression
    '''
    # add a constant to the data
    x = sm.add_constant(x)

    # create the model object
    model = sm.OLS(y, x)

    # and return the fit model
    return model.fit()

# the file name of the dataset
r_filename = '../../Data/Chapter6/power_plant_dataset_pc.csv'

# read the data
csv_read = pd.read_csv(r_filename)
```

```
# select the names of columns
dependent = csv_read.columns[-1]
independent_reduced = [
    col
    for col
    in csv_read.columns
    if col.startswith('p')
]

independent = [
    col
    for col
    in csv_read.columns
    if      col not in independent_reduced
        and col not in dependent
]

# split into independent and dependent features
x = csv_read[independent]
y = csv_read[dependent]

# estimate the model using all variables (without PC)
regressor = regression_ols(x,y)
print(regressor.summary())
```

How it works...

First, as usual, we read in the required data and select the dependent and independent variables.

Then, we estimate the model. First, we add a constant to our dataset; this is nothing more than adding a column of ones. We then create the regressor model using `.OLS(...)`; this method, in contrast to `Scikit`'s equivalent, accepts the dependent and independent variables. Finally, we fit the model and return it. Then, we print the model summary (something that is missing in `Scikit`'s implementation of linear regression). We abbreviated the table to save space:

```
The method regression_ols took 0.03 sec to run.
                          OLS Regression Results
==============================================================================
Dep. Variable:      net_generation_MWh   R-squared:                    0.997
Model:                             OLS   Adj. R-squared:               0.997
Method:                  Least Squares   F-statistic:               4.641e+04
Date:                 Fri, 18 Mar 2016   Prob (F-statistic):            0.00
Time:                         20:25:42   Log-Likelihood:               17787.
No. Observations:                 4494   AIC:                      -3.552e+04
Df Residuals:                     4465   BIC:                      -3.533e+04
Df Model:                           28
Covariance Type:             nonrobust
==============================================================================
                      coef    std err          t      P>|t|      [95.0% Conf. Int.]
------------------------------------------------------------------------------
const              -0.0021      0.000    -18.943      0.000      -0.002     -0.002
fuel_aer_NG         0.0024      0.000      8.143      0.000       0.002      0.003
fuel_aer_DFO        0.0030      0.000      9.272      0.000       0.002      0.004
fuel_aer_HYC        0.0013   9.81e-05     13.060      0.000       0.001      0.001
fuel_aer_SUN        0.0043      0.001      4.084      0.000       0.002      0.006
fuel_aer_WND        0.0015      0.000     14.140      0.000       0.001      0.002
fuel_aer_COL       -0.0028      0.000     -7.838      0.000      -0.003     -0.002
...
state_IA           -0.0007      0.000     -1.799      0.072      -0.001   5.87e-05
state_IL           -0.0015      0.000     -4.632      0.000      -0.002     -0.001
state_OH            0.0003      0.000      0.845      0.398      -0.000      0.001
state_GA            0.0005      0.000      1.235      0.217      -0.000      0.001
state_WA            0.0008      0.000      1.943      0.052   -7.08e-06      0.002
total_fuel_cons    -0.0679      0.002    -31.755      0.000      -0.072     -0.064
total_fuel_cons_mmbtu  0.9881   0.001    732.116      0.000       0.986      0.991
==============================================================================
Omnibus:                      2868.689   Durbin-Watson:                1.926
Prob(Omnibus):                   0.000   Jarque-Bera (JB):       1799556.652
Skew:                           -1.698   Prob(JB):                      0.00
Kurtosis:                      100.974   Cond. No.                  4.54e+15
==============================================================================

Warnings:
[1] Standard Errors assume that the covariance matrix of the errors is correctly specified
[2] The smallest eigenvalue is 3.13e-28. This might indicate that there are
strong multicollinearity problems or that the design matrix is singular.
```

You can see clearly that most of the fuel type, mover, and state variables are not statistically significant: the `p-value` for these variables is greater than `0.05`. These variables can be safely removed from the model without losing much of the accuracy. Note, also, the warning number 2 telling us that there is a strong multicollinearity problem with our dataset—something that we already know.

The R^2 and high statistical significance of `total_fuel_cons` and `total_fuel_cons_mmbtu` confirm our earlier findings: these two variables are essentially the only ones that determine the power plant output. This finding also suggests that there is virtually no difference in what type of fuel we use: we get the same amount of electric energy from the same amount of input energy irrespective of what fuel type we use. We need to remember that a power plant is essentially a power converter: it converts one type of energy to another. This raises an important question—why use an energy source like fossil fuels instead of renewables?

Let's check whether our model is equally effective if we keep only these two variables:

```
# remove insignificant variables
significant = ['total_fuel_cons', 'total_fuel_cons_mmbtu']
x_red = x[significant]

# estimate the model with limited number of variables
regressor = regression_ols(x_red,y)
print(regressor.summary())
```

The model's performance did not deteriorate at all:

```
                          OLS Regression Results
==============================================================================
Dep. Variable:       net_generation_MWh   R-squared:                       0.996
Model:                              OLS   Adj. R-squared:                  0.996
Method:                   Least Squares   F-statistic:                 5.498e+05
Date:                Fri, 18 Mar 2016   Prob (F-statistic):               0.00
Time:                        20:25:42   Log-Likelihood:                  17400.
No. Observations:                4494   AIC:                         -3.479e+04
Df Residuals:                    4491   BIC:                         -3.478e+04
Df Model:                           2
Covariance Type:            nonrobust
==============================================================================
                          coef    std err          t      P>|t|      [95.0% Conf. Int.]
------------------------------------------------------------------------------
const                  -0.0005    7.9e-05     -5.822      0.000      -0.001      -0.000
total_fuel_cons        -0.0528      0.002    -29.063      0.000      -0.056      -0.049
total_fuel_cons_mmbtu   0.9636      0.001    975.693      0.000       0.962       0.966
==============================================================================
Omnibus:                     1908.631   Durbin-Watson:                    1.794
Prob(Omnibus):                  0.000   Jarque-Bera (JB):          1599062.386
Skew:                          -0.484   Prob(JB):                          0.00
Kurtosis:                      95.406   Cond. No.                          24.9
==============================================================================

Warnings:
[1] Standard Errors assume that the covariance matrix of the errors is correctly specified.
```

As you can see, the R^2 dropped from `0.997` to `0.996`, a difference that is negligible. Note also that the constant, even though statistically significant, has almost no influence on the final output. In fact, the only variable worth keeping in the model is `total_fuel_cons_mmbtu`, as it has the strongest (and linear) relationship with the amount of electric energy generated.

There's more...

We can also use `MLPY` to estimate the OLS model (the `regression_ols_alternative.py` file):

```python
import mlpy as ml

@hlp.timeit
def regression_linear(x,y):
    '''
        Estimate a linear regression
    '''
    # create the model object
    ols = ml.OLS()

    # estimate the model
    ols.learn(x, y)

    # and return the fit model
    return ols

# remove insignificant variables
significant = ['total_fuel_cons_mmbtu']
x = csv_read[significant]
y = csv_read[csv_read.columns[-1]]

# estimate the model using all variables (without PC)
regressor = regression_linear(x,y)

# predict the output
predicted = regressor.pred(x)

# and calculate the R^2
score = hlp.get_score(y, predicted)
print('R2: ', score)
```

The `MLPY`'s `.OLS()` can build only simple linear models. We will use our most significant variable, `total_fuel_cons_mmbtu`. As the `MLPY` does not provide any information about the performance of our model, we decided to develop the scoring by hand (the `helper.py` file):

```python
def get_score(y, predicted):
    '''
        Method to calculate R^2
    '''
    # calculate the mean of actuals
```

```
mean_y = y.mean()

# calculate the total sum of squares and residual
# sum of squares
sum_of_square_total = np.sum((y - mean_y)**2)
sum_of_square_resid = np.sum((y - predicted)**2)

return 1 - sum_of_square_resid / sum_of_square_total
```

The method follows the standard definition of the coefficient of determination: it is a ratio of the residual sum of squares to the total sum of squares subtracted from 1. The total sum of squares is a sum of all the squared differences between each observation in the dependent variable and their mean. The residual sum of squares is the sum of squared differences between the actual observed value of the dependent variable and the one output by the model.

The get_score(...) method required two input parameters: the actual observed values of the dependent variable and the ones predicted by the model.

For this model, we got the following:

```
R2:   0.9951670803634637
```

See also

Check this website for another approach to OLS:

```
http://www.datarobot.com/blog/ordinary-least-squares-in-python/
```

Estimating the output of an electric plant using CART

Previously, we used decision trees to classify our bank contact calls (refer to the *Classifying calls with decision trees* recipe from *Chapter 3, Classification Techniques*). Classification and regression trees are the equivalent methods applied to regression problems.

Getting ready

To execute this recipe, you need pandas and Scikit. No other prerequisites are required.

How to do it...

Estimating CART with `Scikit` is extremely easy (the `regression_cart.py` file):

```python
import sklearn.tree as sk

@hlp.timeit
def regression_cart(x,y):
    '''
        Estimate a CART regressor
    '''
    # create the regressor object
    cart = sk.DecisionTreeRegressor(min_samples_split=80,
        max_features="auto", random_state=666666,
        max_depth=5)

    # estimate the model
    cart.fit(x,y)

    # return the object
    return cart
```

How it works...

As with all the other recipes, we first load the data and extract the dependent variable `y` and independent variables `x`. We also keep the independent variables in the form of principal components separately in `x_red`.

To estimate our model, first, as with all the models we estimate using `Scikit`, we create the model object. The `.DecisionTreeRegressor(...)` accepts a multitude of parameters. The `min_samples_split` determines the minimum number of samples that need to be present in the node so that the tree can perform the split. The `max_features` determines the maximum number of variables the tree is allowed to use; the `auto` parameter means, however, that the model can use all the independent variables if necessary. The `random_state` determines the initial random state of the model and `max_depth` determines the maximum level of levels allowed in the tree.

Then, we fit the data and return the estimated model. Let's see how the model performs:

```python
print('R2: ', regressor.score(x,y))
```

This looks like it attains almost the same R^2 as our previous models:

R2: 0.959545513651

The beauty of decision trees is that the insignificant independent variables are not used in the model at all. Let's see which ones are actually significant:

```
for counter, (nm, label) \
    in enumerate(
        zip(x.columns, regressor.feature_importances_)
    ):
    print("{0}. {1}: {2}".format(counter, nm, label))
```

We loop through `feature_importances_`, an attribute of `regressor` (`.DecisionTreeRegressor(...)`) and check what is the name of the corresponding feature using `zip(...)`. The following list is abbreviated:

```
0. fuel_aer_NG: 0.0
1. fuel_aer_DFO: 0.0
...
29. state_GA: 0.0
30. state_WA: 0.0
31. total_fuel_cons: 0.0
32. total_fuel_cons_mmbtu: 1.0
```

As you can see, the model picks up only `total_fuel_cons_mmbtu` as significant, accounting for 100% of the model's predictive power; all the other variables have no impact on the model's performance.

This can be confirmed by looking at the resulting decision tree:

```
sk.export_graphviz(regressor,
    out_file='../../Data/Chapter6/CART/tree.dot')
```

To convert the `.dot` file to something readable, you can use the `dot` command from the command line. First, navigate to the folder (assuming that you are in the `Codes/Chapter6` folder):

cd ../../Data/Chapter6/CART

Then, issue the following command:

dot -Tpdf tree.dot -o tree.pdf

The resulting tree should have only one decision variable:

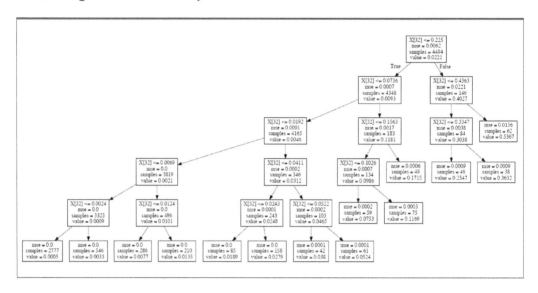

The X[32], if you look it up in the list of columns, is `total_fuel_cons_mmbtu` as expected.

There's more...

Let's check now how would a model with only principal components perform:

```
regressor_red = regression_cart(x_red,y)
```

The R² attained by this model is borderline acceptable:

```
R2:    0.703565587694
```

What is actually quite surprising (if you remember the percentages of the variance explained by each of the principal components) that the third PC is the most important of all of them, and the second one is not being used at all:

```
0. p_0: 0.1727325749575984
1. p_1: 0.08687858306122068
2. p_2: 0.0
3. p_3: 0.542957699724175
4. p_4: 0.1974311422570059
```

> The problem with using principal components (in general) is the fact that one cannot interpret the results directly and deriving the actual meaning of them is somewhat convoluted (https://onlinecourses.science.psu.edu/stat505/node/54).

The resulting tree is not much more complicated than the preceding one:

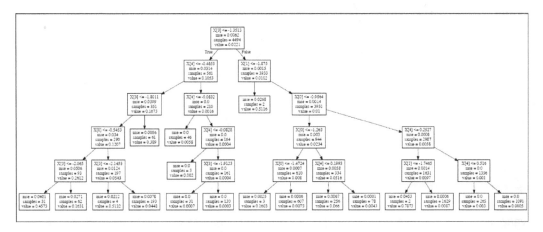

See also

A good read on CART can be found here:

`http://www.stat.wisc.edu/~loh/treeprogs/guide/wires11.pdf`

Employing the kNN model in a regression problem

Although used predominantly to solve classification problems, the k-Nearest Neighbors model that we saw in *Chapter 3, Classification Techniques*, can also be used in regression models. This recipe will teach you how it can be applied.

Getting ready

To execute this recipe, you will need `pandas` and `Scikit`. No other prerequisites are required.

How to do it...

Again, using `Scikit` to estimate this model is extremely simple (the `regression_knn.py` file):

```
import sklearn.neighbors as nb

@hlp.timeit
def regression_kNN(x,y):
    '''
        Build the kNN classifier
```

```
'''

# create the classifier object
knn = nb.KNeighborsRegressor(n_neighbors=80,
    algorithm='kd_tree', n_jobs=-1)

# fit the data
knn.fit(x,y)

# return the classifier
return knn
```

How it works...

First, we read the data in and split it into the dependent variable `y` and independent variables `x_sig`; we are selecting only the significant variables that we found earlier, `total_fuel_cons` and `total_fuel_cons_mmbtu`. We will also test a model built with principal components only; for this, we will use `x_principal`.

To build the kNN regression model, we will use `Scikit`'s `.KNeighborsRegressor(...)` method. We specify `n_neighbors` to `80` and `kd_tree` as the algorithm to be used.

 Check here for an explanation on how kd-Trees are formed:
`http://www.alglib.net/other/nearestneighbors.php`

Specifying the number of jobs to `-1` tells the model to spin up as many parallel jobs as there are cores in your processor.

The model that we estimated attained the R^2 score of `0.94`.

However, a question arises: how sensitive is the performance of the model to the data? To test this, we can use the `cross_validation` module of `Scikit`:

```
import sklearn.cross_validation as cv

# test the sensitivity of R2
scores = cv.cross_val_score(regressor, x_sig, y, cv=100)
print('Expected R2: {0:.2f} (+/-{1:.2f})'\
    .format(scores.mean(), scores.std()**2))
```

First, we load the module. Then, we use the `.cross_val_score(...)` method to test our model. The first parameter is the model to be tested, second parameter is the set of our independent variables, and third one specifies the dependent variable. The last parameter passed, `cv`, specifies the number of folds: the number of partitions that the original dataset is randomly divided into. Normally, only one of the partitions would be used in cross-validation, with the remaining used as a training dataset.

Here is what we got:

```
R2:   0.943391131036
Expected R2: 0.91 (+/- 0.03)
```

So, we got R^2 of `0.94` from our run. However, the cross-validation suggests that the expected value is `0.91` with a standard deviation of `0.03` (so our run is within one standard deviation).

Let's test the model with principal components only:

```
# estimate the model using Principal Components only
regressor_principal = regression_kNN(x_principal,y)

print('R2: ', regressor_principal.score(x_principal,y))

# test the sensitivity of R2
scores = cv.cross_val_score(regressor_principal,
    x_principal, y, cv=100)
print('Expected R2: {0:.2f} (+/- {1:.2f})'\
    .format(scores.mean(), scores.std()**2))
```

What we got in return might be quite surprising (especially if you were taught that the R^2 can only take values between `0` and `1`):

```
R2:   0.460588198514
Expected R2: -18.53 (+/- 14426.26)
```

So, for this run, we got R^2 of `0.46`. However, if you check the expected R^2, the value is negative (and the standard deviation is extremely high).

How can we get a negative value of R^2? Consider the following example: we have a data that more or less follows the relationship of $y=x^2$. Now, let's assume that some model (capable of estimating only linear relationships) gives us a prediction that follows the `y_pred=3*x` function.

For such a predictor, R^2 would be around -0.4. If you analyze how R^2 is calculated, you will realize that R^2 can get negative in certain situations.

 Check out this answer from StackOverflow:
`http://stats.stackexchange.com/questions/12900/when-is-r-squared-negative`

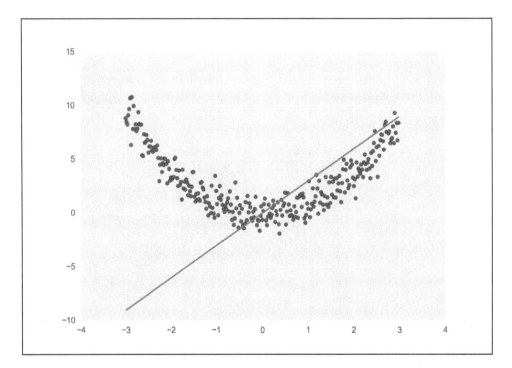

Applying the Random Forest model to a regression analysis

Random Forest, similar to decision trees, can also be applied to solving regression problems. We used them previously to classify calls (refer to the *Predicting subscribers with random tree forests* recipe in *Chapter 3, Classification Techniques*). Here, we will use Random Forest to predict the output of a power plant.

Getting ready

To execute this, you will need `pandas`, `NumPy`, and `Scikit`. No other prerequisites are required.

How to do it...

The Random Forests are part of the ensemble types of models. This example borrows the code-shell that we presented in *Chapter 3, Classification Techniques* (the `regression_randomForest.py` file):

```python
import sys
sys.path.append('..')

# the rest of the imports
import helper as hlp
import pandas as pd
import numpy as np
import sklearn.ensemble as en
import sklearn.cross_validation as cv

@hlp.timeit
def regression_rf(x,y):
    '''
        Estimate a random forest regressor
    '''
    # create the regressor object
    random_forest = en.RandomForestRegressor(
        min_samples_split=80, random_state=666,
        max_depth=5, n_estimators=10)

    # estimate the model
    random_forest.fit(x,y)

    # return the object
    return random_forest
```

How it works...

As always, we will first read in the data and split it into independent and dependent variables. These will then be used to estimate our models.

The `.RandomForestRegressor(...)` method creates the model object. The `min_samples_split`, just like with decision trees, determines the minimum number of samples in the node to allow further splits. The `random_state` specifies the initial random state of the model and `max_dept` defines the maximum depth of the tree (how many levels). The last parameter, `n_estimators`, instructs the model on how many trees to estimate in the forest.

Once estimated, we print out the R^2 score, test its sensitivity to input, and output the list of variables with their related importance:

```
# print out the results
print('R: ', regressor.score(x,y))

# test the sensitivity of R2
scores = cv.cross_val_score(regressor, x, y, cv=100)
print('Expected R2: {0:.2f} (+/- {1:.2f})'\
    .format(scores.mean(), scores.std()**2))

# print features importance
for counter, (nm, label) \
    in enumerate(
        zip(x.columns, regressor.feature_importances_)
    ):
    print("{0}. {1}: {2}".format(counter, nm, label))
```

The preceding code should generate an output similar to the following one (abbreviated):

```
The method regression_rf took 0.05 sec to run.
R:   0.970459248524
Expected R2: 0.82 (+/- 0.21)
0. fuel_aer_NG: 0.0
1. fuel_aer_DFO: 0.0
2. fuel_aer_HYC: 0.0
3. fuel_aer_SUN: 0.0
4. fuel_aer_WND: 0.0
5. fuel_aer_COL: 0.0
6. fuel_aer_MLG: 0.0
7. fuel_aer_NUC: 0.0
8. mover_CT: 0.0
9. mover_GT: 0.0
10. mover_HY: 0.0
11. mover_IC: 0.0
12. mover_PV: 0.0
13. mover_ST: 1.9909417055476134e-05
14. mover_WT: 0.0
15. state_CA: 0.0
16. state_TX: 0.0
17. state_NY: 0.0
18. state_FL: 0.0
19. state_MN: 0.0
20. state_MI: 0.0
21. state_NC: 0.0
22. state_PA: 0.0
23. state_MA: 0.0
24. state_WI: 0.0
25. state_NJ: 0.0
26. state_IA: 0.0
27. state_IL: 0.0
28. state_OH: 0.0
29. state_GA: 0.0
30. state_WA: 0.0
31. total_fuel_cons: 0.0
32. total_fuel_cons_mmbtu: 0.9999800905829446
```

Once again, we can see that `total_fuel_cons_mmbtu` dominates the model (as expected). The `mover_ST` shows up but its importance is next to nothing so we can remove it.

Building the model with only `total_fuel_cons_mmbtu` leads to the following results:

```
The method regression_rf took 0.02 sec to run.
R:   0.970432620578
Expected R2: 0.81 (+/- 0.22)
0. total_fuel_cons_mmbtu: 1.0
```

As you can see, the difference in R^2 and sensitivity is negligible.

Gauging the amount of electricity a plant can produce using SVMs

Support Vector Machines (**SVMs**) have gained popularity over the years. These very powerful models use kernel tricks to model even the most complicated relationships between the dependent and independent variables.

In this recipe, using artificially generated data, we will show the real power of SVMs.

Getting ready

To execute this recipe, you will need `pandas`, `NumPy`, `Scikit`, and `Matplotlib`. No other prerequisites are required.

How to do it...

In this recipe, we will test SVM for regression with four different kernels (the `regression_svm.py` file):

```python
import sys
sys.path.append('..')

# the rest of the imports
import helper as hlp
import pandas as pd
import numpy as np
import sklearn.svm as sv
import matplotlib.pyplot as plt

@hlp.timeit
```

```
def regression_svm(x, y, **kw_params):
    '''
        Estimate a SVM regressor
    '''
    # create the regressor object
    svm = sv.SVR(**kw_params)

    # estimate the model
    svm.fit(x,y)

    # return the object
    return svm

# simulated dataset
x = np.arange(-2, 2, 0.004)
errors = np.random.normal(0, 0.5, size=len(x))
y = 0.8 * x**4 - 2 * x**2 +  errors

# reshape the x array so its in a column form
x_reg = x.reshape(-1, 1)

models_to_test = [
    {'kernel': 'linear'},
    {'kernel': 'poly','gamma': 0.5, 'C': 0.5, 'degree': 4},
    {'kernel': 'poly','gamma': 0.5, 'C': 0.5, 'degree': 6},
    {'kernel': 'rbf','gamma': 0.5, 'C': 0.5}
]
```

How it works...

First, we load all the necessary modules. Next, we simulate the dataset. We start by generating the x vector using the `.arange(...)` method of `NumPy`; our dataset will start at `-2` and extend to 2 with equal steps of `0.004`. We then introduce some errors: the `.random.normal(...)` method returns the same number of samples as the length of x drawn from a normal distribution with `0` mean and `0.5` standard deviation. Our dependent variable is specified as follows: $y = 0.8x^4 - 2x^2 + errors$.

For the purpose of estimating the model, we need to have our x independent variable in a columnar form; we use the `.reshape(...)` method to change the shape of our x.

The `models_to_test` list contains a list of four keyword parameter specifications for our `regression_svm(...)` method; we will estimate one SVM model with a linear kernel, two SWM models with polynomial kernels (with a degree of 4 and degree of 6), and one model with an RBF kernel.

The `regression_svm(...)` method follows the same logic as all our other model estimation methods: first, we create the model object, then we estimate the model, and finally, we return the estimated model.

In this recipe, we will produce a chart to illustrate the model predictions.

First, let's create the figure and adjust its frame:

```
plt.figure(figsize=(len(models_to_test) * 2 + 3, 9.5))
plt.subplots_adjust(left=.05, right=.95,
    bottom=.05, top=.96, wspace=.1, hspace=.15)
```

First, we create a figure with specified size and then adjust the subplots: we alter the `left` and `right`, `bottom` and `top` margins for each subplot and adjust the space in between the subplots, `wspace` and `hspace`.

Now, we will loop through all the models that we want to build and add them to the figure:

```
for i, model in enumerate(models_to_test):
    # estimate the model
    regressor = regression_svm(x_reg, y, **model)

    # score
    score = regressor.score(x_reg, y)

    # plot the chart
    plt.subplot(2, 2, i + 1)
    if model['kernel'] == 'poly':
        plt.title('Kernel: {0}, deg: {1}'\
            .format(model['kernel'], model['degree']))
    else:
        plt.title('Kernel: {0}'.format(model['kernel']))
    plt.scatter(x, y)
    plt.plot(x, regressor.predict(x_reg), color='r')
    plt.ylim([-4, 8])
    plt.text(.9, .9, ('R^2: {0:.2f}'.format(score)),
                transform=plt.gca().transAxes, size=15,
                horizontalalignment='right')
```

First, we estimate our model by calling the `regression_svm(...)` method. Once the model is estimated, we calculate the R^2 score.

We then specify (on the 2-by-2 grid) where we are going to put our current model's plot using `.subplot(...)` and taking the current iteration `i` increased by 1; the numbering of subplots in the chart starts with 1 while the first element of a list has an index of 0. If the kernel is polynomial, the title of our chart should reflect the degree; otherwise, the kernel name in the title will suffice.

Next, we create a scatter plot with our original data: x and y. Following this, we plot the curve with what the regressor predicts; in order to make it stand out, we change the color to red. We want to keep the y axis homogeneous across all the charts so we specify the limits to be between -4 and 8.

Lastly, we put the R^2 value in the top right corner; this is achieved by specifying the coordinates (.9, .9) and using transform. We push the text to be right-aligned and specify its size to 15.

Finally, we save the final chart to a file:

```
plt.savefig('../../data/Chapter6/charts/regression_svm.png',
    dpi= 300)
```

The resulting chart shows how well the RBF kernel handles highly nonlinear data. Once again, we see a negative value for R^2 for the model with the linear kernel. You can notice that even the polynomial kernels do not represent the data well enough:

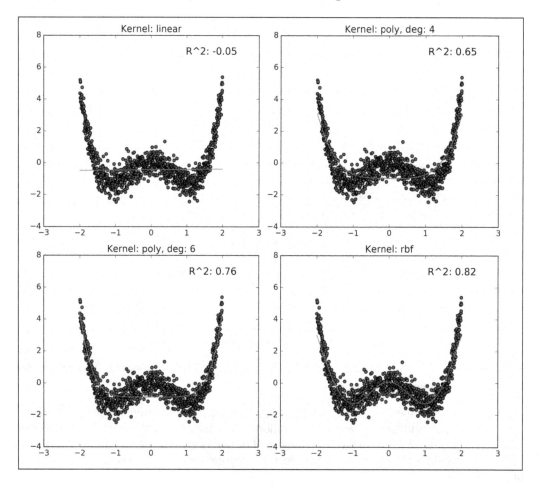

There's more...

The main difficulty in estimating SVMs with RBF kernels is deciding what values should be chosen for the C (the error penalty parameter) and gamma (the RBF kernel parameter that controls the sensitivity to differences in feature vectors) parameters.

To this end, we can use a grid search, which might be a highly computationally-intensive undertaking. Instead, we will use Optunity, a library with various optimizers for many environments.

 Optunity can be found here:
http://optunity.readthedocs.org/en/latest/

Optunity is not supported directly by Anaconda so we need to install it ourselves. In your command line (on a Unix-like system), issue the following commands (assuming that you have a Downloads folder in your home folder):

```
cd ~/Downloads
git clone https://github.com/claesenm/optunity.git
cd optunity
python setup.py install
```

If everything goes well, you should not receive any errors.

Now that we have installed Optunity, let's turn to our code (the regression_svm_alternative.py file). I am going to skip the usual imports here; you can look up the full import list in the file:

```
import optunity as opt
import optunity.metrics as mt

# simulated dataset
x = np.arange(-2, 2, 0.002)
errors = np.random.normal(0, 0.5, size=len(x))
y = 0.8 * x**4 - 2 * x**2 +  errors

# reshape the x array so its in a column form
x_reg = x.reshape(-1, 1)

@opt.cross_validated(x=x_reg, y=y, num_folds=10, num_iter=5)
def regression_svm(
    x_train, y_train, x_test, y_test, logC, logGamma):
    '''

        Estimate a SVM regressor
```

```
'''

    # create the regressor object
    svm = sv.SVR(kernel='rbf',
        C=0.1 * logC, gamma=0.1 * logGamma)

    # estimate the model
    svm.fit(x_train,y_train)

    # decision function
    decision_values = svm.decision_function(x_test)

    # return the object
    return mt.roc_auc(y_test, decision_values)

# find the optimal values of C and gamma
hps, _, _ = opt.maximize(regression_svm, num_evals=10,
    logC=[3, 10], logGamma=[3, 20])
```

First, we create the simulated dataset in the same fashion as in the previous example.

Then, we decorate our `regression_svm(...)` method with a method from optunity. The `.cross_validated(...)` method accepts a number of parameters: the x points to the set of independent variables and y is the dependent variable.

The `num_folds` specifies the number of cross-validation folds. In our case, we will use 1/10 of our sample for cross-validation and 9/10 as training. The `num_iter` defines how many iterations to use in order to cross-validate.

Note that the definition of our `regression_svm(...)` method changed as well. The `cross_validated(...)` method injects the training x and y samples first, followed by testing x and y and all the other parameters we specify: in our case, `logC` and `logGamma`. The latter two parameters are the ones that we want to find the optimal values for.

The method starts by estimating the model: first, we create the model object with specified parameters and then fit our data to it. The `decision_function(...)` method returns a list of distances between each element of x to the separating hyperplane. We return the value of the area under the receiver operating characteristic value. (Refer to *Testing and comparing the models* recipe from *Chapter 3, Classification Techniques* for more details on ROC.)

Using the `maximize(...)` method provided by optunity, we find the optimal values of C and gamma. The first parameter is the function that we want to maximize, `regression_svm(...)` in our case. The `num_evals` specifies the number of permitted function evaluations. The last two specify the bounds for the `logC` and `logGamma` search space.

Once optimized, we can use the found values to estimate the final model and save the chart:

```
# and the values are...
print('The optimal values are: C - {0:.2f}, gamma - {1:.2f}'\
    .format(0.1 * hps['logC'], 0.1 * hps['logGamma']))

# estimate the model with optimal values
regressor = sv.SVR(kernel='rbf',
            C=0.1 * hps['logC'],
            gamma=0.1 * hps['logGamma'])\
        .fit(x_reg, y)

# predict the output
predicted = regressor.predict(x_reg)

# and calculate the R^2
score = hlp.get_score(y, predicted)
print('R2: ', score)

# plot the chart
plt.scatter(x, y)
plt.plot(x, predicted, color='r')
plt.title('Kernel: RBF')
plt.ylim([-4, 8])
plt.text(.9, .9, ('R^2: {0:.2f}'.format(score)),
            transform=plt.gca().transAxes, size=15,
            horizontalalignment='right')

plt.savefig(
    '../../data/Chapter6/charts/regression_svm_alt.png',
    dpi=300
)
```

We have already covered most of this code before so we will not be repeating ourselves here.

The values found should be as follows:

```
The optimal values are: C - 0.68, gamma - 1.31
R2:  0.8605360224
```

Compared with our previous example, R^2 increased; now, the regression line fits the data even better:

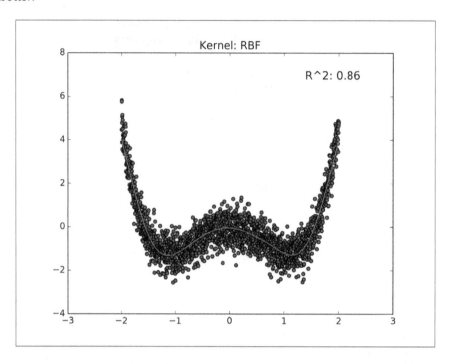

See also

I strongly recommend getting familiar with `Optunity`:

`http://optunity.readthedocs.org/en/latest/`

Training a Neural Network to predict the output of a power plant

Neural networks are really powerful regressors, capable of fitting almost any arbitrary data. In this example, however, we will use a simple model to fit the data as we already know our data exerts a linear relationship.

Getting ready

To execute this recipe, you will need `pandas` and `PyBrain`. No other prerequisites are required.

How to do it...

We specify the `fitANN(...)` method in a very similar way as we did in *Chapter 3, Classification Techniques* (the `regression_ann.py` file). For brevity, some of the import statements have been removed:

```
import pybrain.structure as st
import pybrain.supervised.trainers as tr
import pybrain.tools.shortcuts as pb

@hlp.timeit
def fitANN(data):
    '''
        Build a neural network regressor[1]
    '''

    # create the regressor object
    ann = pb.buildNetwork(inputs_cnt,
        inputs_cnt * 3,
        target_cnt,
        hiddenclass=st.TanhLayer,
        outclass=st.LinearLayer,
        bias=True
    )

    # create the trainer object
    trainer = tr.BackpropTrainer(ann, data,
        verbose=True, batchlearning=False)

    # and train the network
    trainer.trainUntilConvergence(maxEpochs=50, verbose=True,
        continueEpochs=2, validationProportion=0.25)

    # and return the regressor
    return ann
```

How it works...

We start with reading in the data, splitting the data, and creating the datasets for the ANN training:

```
# split the data into training and testing
train_x, train_y, \
test_x,  test_y, \
labels = hlp.split_data(csv_read,
    y='net_generation_MWh', x=['total_fuel_cons_mmbtu'])
```

```
# create the ANN training and testing datasets
training = hlp.prepareANNDataset((train_x, train_y),
    prob='regression')
testing  = hlp.prepareANNDataset((test_x, test_y),
    prob='regression')
```

We are only using the `total_fuel_cons_mmbtu` variable as our independent feature. Note that we use a new parameter for the `.prepareANNDataset(...)` method.

The prob (standing for problem) specifies that we want a dataset to solve a regression problem; for the classification problem, we needed to have the class indicator and its complement to 1, `abs(item - 1)`. Refer to the *Employing neural networks to classify calls* recipe in *Chapter 3, Classification Techniques*.

We are now ready to fit the model:

```
# train the model
classifier = fitANN(training)

# predict the output for the unseen data
predicted = classifier.activateOnDataset(testing)

# and calculate the R^2
score = hlp.get_score(test_y, predicted[:, 0])
print('R2: ', score)
```

The `fitANN(...)` method first creates the neural network. Our neural network has one input neuron, three neurons in the hidden layer (with `tanh` activation functions), and one output neuron (with a linear activation function).

The trainer uses online training (`batchlearning=False`), as we did in *Chapter 3, Classification Techniques*. We specified `maxEpochs` to `50` so that if our network did not minimize the error by then, we will stop anyway. We will use a quarter of our dataset for validation purposes.

After the training, we test the network by predicting the output for the unseen data and calculating R^2:

```
R2: 0.9871850894104638
```

As you can see, our neural network performs as good as most of our other models.

See also

Check out this example of how to train neural networks with `PyBrain`:

http://fastml.com/pybrain-a-simple-neural-networks-library-in-python/

7
Time Series Techniques

In this chapter, we will cover various techniques of handling, analyzing, and predicting the future with time series data. You will learn the following recipes:

- ▶ Handling date objects in Python
- ▶ Understanding time series data
- ▶ Smoothing and transforming the observations
- ▶ Filtering the time series data
- ▶ Removing trend and seasonality
- ▶ Forecasting the future with ARMA and ARIMA models

Introduction

Time series can be found everywhere; if you analyze the stock market, sunspot occurrences, or river flows, you are observing phenomena that are stretched in time. It is almost inevitable that any data scientist throughout his or her career will deal with time series data at some point. In this chapter, we will see various techniques of handling, analyzing, and building models for time series.

The datasets for this chapter come from the web archive of river flows, which can be accessed here:

`http://ftp.uni-bayreuth.de/math/statlib/datasets/riverflow`

The archive is essentially a shell script that we processed to create the datasets for this chapter. In order to create the raw files from the archive, you can use Cygwin (on Windows) or Terminal on Mac/Linux and execute the following command (assuming that you save the archive in `riverflows.webarchive`):

```
sh riverflow.webarchive
```

Handling date objects in Python

Time series is data that is sampled at certain intervals over time, for example, recording the speed of a car every second. Given such data, we can easily estimate either the distance traveled (by summing the observations and dividing the sum by 3,600) or acceleration of the car (by calculating the differences between two consecutive observations). Managing time series data with `pandas` is straightforward.

Getting ready

To execute this recipe, you will need `pandas`, `NumPy`, and `Matplotlib`. No other prerequisites are necessary.

How to do it...

From the web archive, we cleaned up and transformed two datasets: water flow for the American River (`http://www.theamericanriver.com`) and for the Columbia River (`http://www.ecy.wa.gov/programs/wr/cwp/cwpfactmap.html`). With `pandas`, it is extremely easy to read the time series dataset (the `ts_handlingData.py` file):

```python
import numpy as np
import pandas as pd
import pandas.tseries.offsets as ofst

# files we'll be working with
files = ['american.csv', 'columbia.csv']

# folder with data
data_folder = '../../Data/Chapter07/'

# read the data
american = pd.read_csv(data_folder + files[0],
    index_col=0, parse_dates=[0],
    header=0, names=['','american_flow'])

columbia = pd.read_csv(data_folder + files[1],
    index_col=0, parse_dates=[0],
    header=0, names=['','columbia_flow'])

# combine the datasets
riverFlows = american.combine_first(columbia)

# periods aren't equal in the two datasets so find the overlap
```

```
# find the first month where the flow is missing for american
idx_american = riverFlows \
    .index[riverFlows['american_flow'].apply(np.isnan)].min()

# find the last month where the flow is missing for columbia
idx_columbia = riverFlows \
    .index[riverFlows['colum_flow'].apply(np.isnan)].max()

# truncate the time series
riverFlows = riverFlows.truncate(
    before=idx_columbia + ofst.DateOffset(months=1),
    after=idx_american - ofst.DateOffset(months=1))

# write the truncated dataset to a file
with open(data_folder + 'combined_flow.csv', 'w') as o:
    o.write(riverFlows.to_csv(ignore_index=True))
```

How it works...

As always, we first import all the necessary modules: pandas and NumPy. We have two files that we will be working with: american.csv and columbia.csv. These are located in data_folder for this chapter.

We will be using the familiar .read_csv(...) method of pandas. First, we read the american.csv file. We instruct the method to treat the first column as an index by specifying index_col=0. In order for pandas to treat that column as a date, we explicitly instruct the .read_csv(...) method to parse_dates in that column.

We will be combining the two files into one dataset. Hence, we decided to alter the names of the columns: we instruct the method that there is no header and provide the names ourselves. Note that the first column does not need any name as it will be converted to the index anyway. We read the data for the Columbia River in the same manner.

After reading the two files, we combine them together. The .combine_first(...) method of pandas takes the first dataset and effectively inserts the columns from the dataset for the Columbia River.

 If we did not have differing names of the columns in our DataFrames, the .combine_first(...) method would use the data from the called DataFrame to fill in the gaps in the caller DataFrame.

The time periods in the files are not equal but they overlap: the data for the American River starts in 1906 and finishes in 1960 while the Columbia River dataset starts in 1933 and goes until the end of 1969. Let's keep the overlapping period; our combined dataset will hold only the data between 1933 and 1960.

First, we find the earliest date for the American River dataset with no data (the `american_ flow` column). We check the index of `riverFlows` and select all the dates for which the value of `american_flow` is not a number; we do this using the `.apply(...)` method and evaluating the elements of the DataFrame with the `.isnan` method from `NumPy`. Having done this, we then select the minimum date from the series.

In contrast, for `columbia_flow`, we find the latest date that the data is not available for. In a similar manner to how we treated the observations for the American River, we first get all the dates where the data is not a number and then select the maximum date from that series.

The `.truncate(...)` method allows us to remove the data from a DataFrame with `DatetimeIndex`.

> `DatetimeIndex` is an immutable array of numbers. Internally represented by large integer numbers, they appear as date-time objects: objects that have both date and time components. To learn more, visit `http://pandas.pydata.org/pandas-docs/ stable/generated/pandas.DatetimeIndex.html`.

We pass two parameters to the `.truncate(...)` method. The `before` parameter defines the cut-off date that all the records before will be discarded, while the `after` parameter specifies the latest possible date to retain the data for.

Our `idx_...` objects hold minimum and maximum dates for which at least one of the columns does not have data. However, if we passed these dates to the `.truncate(...)` method, we would also select one record at each extreme that would have no data. To alleviate this, we offset our dates using the `.DateOffset(...)` method. We shift our dates by one month only.

> If you want to learn more about the `.DateOffset(...)` method, you can look it up here:
>
> `http://pandas.pydata.org/pandas-docs/stable/ timeseries.html#dateoffset-objects`

Lastly, we output the combined dataset to a file. (Refer to *Reading and writing CSV/TSV files with Python* recipe from *Chapter 1, Preparing the Data*, for more information.)

There's more...

Selecting the data from a time series DataFrame is straightforward: you can still use the already known methods of subsetting DataFrames (using indices of elements, for example, `riverFlows[0:10]` or the `.head(...)` method). However, with DataFrames and the `DatetimeIndex` index, you can also pass dates.

```
print(riverFlows['1933':'1934-06'])
```

The preceding command will print all the records between 1933 and June of 1934. Note that you can pass only year even though we have monthly data; `pandas` will pick the first observation for that year if you do this.

Sometimes, you might need to shift your observations in time, for example, your sensor did not adjust its internal clock to daylight saving time and your observations are shifted by an hour or human error caused the observations to be shifted by a year. This can easily be corrected using the `.shfit(...)` method:

```
by_month = riverFlows.shift(1, freq='M')
by_year = riverFlows.shift(12, freq='M')
```

The first parameter to the `.shift(...)` method specifies the number of periods to shift your observations for, and the `freq` parameter defines the frequency (in this example, a month).

Sometimes, you might want to calculate an average (or sum) of your data at a quarter or year end. This can easily be achieved with a `.resample(...)` method:

```
quarter = riverFlows.resample('Q', how='mean')
half = riverFlows.resample('6M', how='mean')
year = riverFlows.resample('A', how='mean')
```

The first parameter determines the frequency: `Q` means quarter end (that is, March, June, September, and December) and `how` specifies what to do with the observations (in our case, we want `mean` as for our data that makes sense, but if you had sales data, you might also specify `sum`).

 A means year end and 6M means half-year end, that is, June and December.

In addition, `pandas` allows one to easily plot any time series data:

```
import matplotlib
import matplotlib.pyplot as plt

# change the font size
matplotlib.rc('xtick', labelsize=9)
```

```
matplotlib.rc('ytick', labelsize=9)
matplotlib.rc('font', size=14)

# colors
colors = ['#FF6600', '#000000', '#29407C', '#660000']

# monthly time series
riverFlows.plot(title='Monthly river flows', color=colors)
plt.savefig(data_folder + '/charts/monthly_riverFlows.png',
    dpi=300)
plt.close()
```

We first load the `Matplotlib` modules and change the font sizes of *x* and *y* axes numbers as well as the title.

 Consult `http://matplotlib.org/users/customizing.html` for a full list of possible `Matplotlib` customizations.

Next, we specify a palette of colors; we will keep this constant for all the charts.

The `.plot(...)` method plots the data. We define the title and allow the method to select the colors from our colors list. Finally, we save the figure to a file and close it so that it frees the memory.

The charts for monthly and quarterly data are presented here:

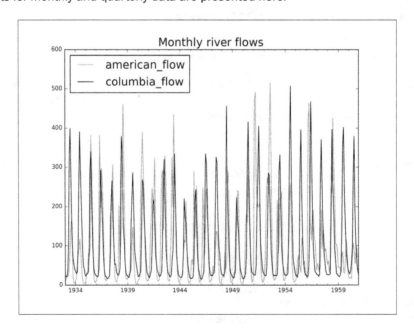

By looking at the monthly flows, you might be tricked into believing that the flows for both of the rivers do not differ much; the quarterly chart already starts to show the differences:

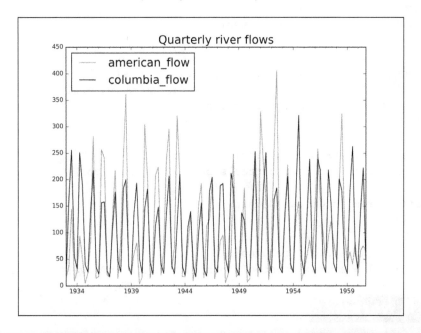

Understanding time series data

One of the fundamental things when dealing with any dataset is to get intimate with it: without understanding what you are dealing with, you cannot build a successful statistical model.

Getting ready

To execute this recipe, you will need pandas, Statsmodels, and Matplotlib. No other prerequisites are required.

How to do it...

One of the fundamental statistics to check for any time series is the **autocorrelation function (ACF)**, **partial autocorrelation function (PACF)**, and spectral density (the ts_timeSeriesFunctions.py file):

```
import statsmodels as sm

# read the data
riverFlows = pd.read_csv(data_folder + 'combined_flow.csv',
```

```
            index_col=0, parse_dates=[0])

    # autocorrelation function
    acf = {}      # to store the results

    for col in riverFlows.columns:
        acf[col] = sm.tsa.stattools.acf(riverFlows[col])

    # partial autocorrelation function
    pacf = {}

    for col in riverFlows.columns:
        pacf[col] = sm.tsa.stattools.pacf(riverFlows[col])

    # periodogram (spectral density)
    sd = {}

    for col in riverFlows.columns:
        sd[col] = sm.tsa.stattools.periodogram(riverFlows[col])
```

How it works...

We omit the module importing and other variables for space-saving reasons here. You should refer to the source file for details.

First, we read in the dataset (which we created in the previous recipe) in a similar manner but we do not specify the names of our variables; instead, we allow the `.read_csv(...)` method to assume these from the first row of the file.

Next, we calculate the ACF. The ACF shows how strongly observations at time t are correlated with observations at time t+lag, where lag defines the distance from the current time. Later in the book (in the *Forecasting the future with ARMA and ARIMA models* recipe), we will use the ACF function to determine the **moving average (MA)** component of the ARMA (or ARIMA) model. Here, we use the `.acf(...)` method of Statsmodels to calculate the ACF for our time series. By default, the nlags parameter of `.acf(...)` is set to 40 and we did not change it.

The PACF can be viewed as a regression of a current observation given the past lags. We will use the plot of the PACF to determine the **auto-regressive (AR)** component in the ARMA (and ARIMA) model. To calculate the PACF, we use the `.pacf(...)` method of Statsmodels.

 If you are interested, you can read more about ACF and PACF online at `https://onlinecourses.science.psu.edu/stat510/node/62`.

Last, we calculate the periodogram, a method that will help us find the fundamental frequency in our data, that is, the dominant frequency of peaks and troughs in our data.

Let's plot it:

```
# plot the data
fig, ax = plt.subplots(2, 3) # 2 rows and 3 columns

# set the size of the figure explicitly
fig.set_size_inches(12, 7)

# plot the charts for American
ax[0, 0].plot(acf['american_flow'], colors[0])
ax[0, 1].plot(pacf['american_flow'],colors[2])
ax[0, 2].plot(sd['american_flow'],  colors[3])
ax[0, 2].yaxis.tick_right() # shows the numbers on the right

# plot the charts for Columbia
ax[1, 0].plot(acf['columbia_flow'], colors[0])
ax[1, 1].plot(pacf['columbia_flow'],colors[1])
ax[1, 2].plot(sd['columbia_flow'],  colors[2])
ax[1, 2].yaxis.tick_right()

# set titles for columns
ax[0, 0].set_title('ACF')
ax[0, 1].set_title('PACF')
ax[0, 2].set_title('Spectral density')

# set titles for rows
ax[0, 0].set_ylabel('American')
ax[1, 0].set_ylabel('Columbia')
```

First, we create a figure that will hold six subplots in two rows and three columns. We set the size of the chart explicitly.

Next, we plot the data. The `ax` is a `NumPy` array with Axes objects; in our example, it is a two-dimensional array. We control where our charts go by specifying the coordinates of the Axes object (where the `0, 0` coordinate is located in the top left corner).

For the last chart in the row, we set the numbers of the *y* axis to be presented on the right of the chart by calling the `.yaxis.tick_right()` method.

We only set the titles for the charts at the top as we plot the ACF, PACF, and Spectral density charts in columns. For the rows, we specify labels for the *y* axis to differentiate between the rivers.

If we plot this, we get the following charts:

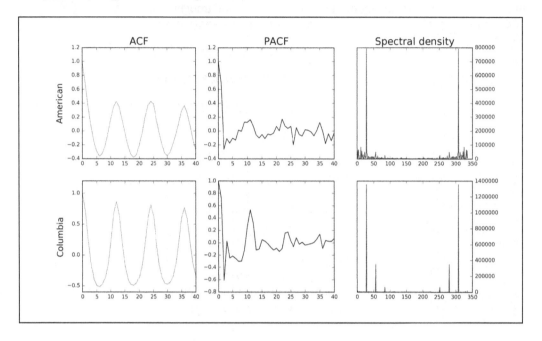

You can clearly see a recurrent pattern in our ACF plots. This suggests that (as expected) our data is seasonal (with a yearly pattern). It also suggests that our process is not stationary.

 A stationary process has mean, variance, and autocorrelation that does not change with time (http://www.itl.nist.gov/ div898/handbook/pmc/section4/pmc442.htm).

The PACF shows that the observation at time t depends strongly on the observations at time $t-1$ and $t-2$ (and to a less extent on the preceding snapshots). Analyzing the spectral density shows the fundamental frequency to be around 29; thus, every 29 months, a fundamental pattern repeats. This is an unexpected pattern as we would presume the sequence to repeat at intervals that are multiples of 12. This might be a function of a number of things: erroneously recorded observations, faulty instrumentation, or fundamentally shifting patterns in a couple of the observed years.

There's more...

The preceding charts for ACF and PACF do not make it easy to see major deviations that cross the confidence intervals: such deviations allow us to easily identify AR and MA components of our time series processes.

However, `Statsmodels` provides a helpful method, `.plot_acf(...)` and `.plot_pacf(...)` (the `ts_timeSeriesFunctions_alternative.py` file):

```python
# plot the data
fig, ax = plt.subplots(2, 2, sharex=True)

# set the size of the figure explicitly
fig.set_size_inches(8, 7)

# plot the charts for american
sm.graphics.tsa.plot_acf(
    riverFlows['american_flow'].squeeze(),
    lags=40, ax=ax[0, 0])

sm.graphics.tsa.plot_pacf(
    riverFlows['american_flow'].squeeze(),
    lags=40, ax=ax[0, 1])

# plot the charts for columbia
sm.graphics.tsa.plot_acf(
    riverFlows['columbia_flow'].squeeze(),
    lags=40, ax=ax[1, 0])

sm.graphics.tsa.plot_pacf(
    riverFlows['columbia_flow'].squeeze(),
    lags=40, ax=ax[1, 1])
```

First, we create the figure that would consist of four subplots (laid out on a 2-by-2 grid). Setting the `sharex` parameter to `True` instructs the Axes object that our plots will all have the same time domain. We also explicitly set the figure size.

The `.plot_acf(...)` and `.plot_pacf(...)` both take the data to plot as their first parameter. The `.squeeze()` method converts a one-column DataFrame into a Series object. Here, we explicitly specify the number of lags for our plots. The `ax` parameter specifies where on the figure grid our chart will be plotted.

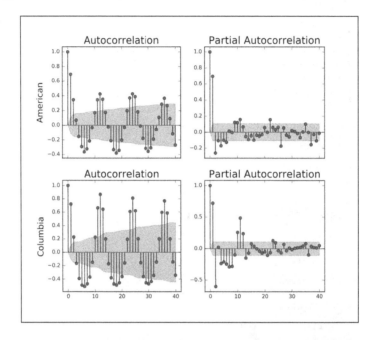

Smoothing and transforming the observations

Smoothing data removes some of the noise from the data and makes the chart appear smooth (hence the name).

Should you do it? For presentation purposes, sure! For modeling purposes, not necessarily—check this argument by William Briggs at `http://wmbriggs.com/post/195/`.

Getting ready

To execute this recipe, you will need `pandas`, `NumPy`, and `Matplotlib`. No other prerequisites are required.

How to do it...

In this recipe, we will see how to calculate the moving averages and transform the data using a logarithmic function (the `ts_smoothing.py` file):

```
ma_transform12   = pd.rolling_mean(riverFlows, window=12)
ma_transformExp  = pd.ewma(riverFlows, span=3)
log_transfrom    = riverFlows.apply(np.log)
```

How it works...

The `.rolling_mean(...)` method of `pandas` calculates a rolling average. It takes the observations between time >t< and >(t + window - 1)<, calculates their average, and uses the calculated figure in place of the observation at time >t<. You can also specify the center parameter that, instead of taking window less one preceding observation, will try to take an equal number of observations on each side of the observation at time `t`.

The `.ewma(...)` method uses an exponentially decaying function to smooth out the data. The `span` parameter controls the influence window—how many observations are still relevant in calculating the weighted average.

 To read more about these techniques, you can study the documentation for `pandas` here:
```
http://pandas.pydata.org/pandas-docs/stable/
computation.html#exponentially-weighted-moment-
functions
```

Transforming the data logarithmically can be beneficial from the computational point of view and sometimes more patterns can be spotted after the log transform. Unlike smoothing, it is a reversible process so we do not lose any accuracy (as long as you do not transform the data further, for example, calculate the rolling average of log-transformed data). Here, we use the `.log` function provided by the `NumPy` library. It takes the observation at time t and window, less one preceding observation, calculates their average, and uses the calculated gure in place of the observation at time `t`

The resulting plot shows you how different techniques transform the original dataset:

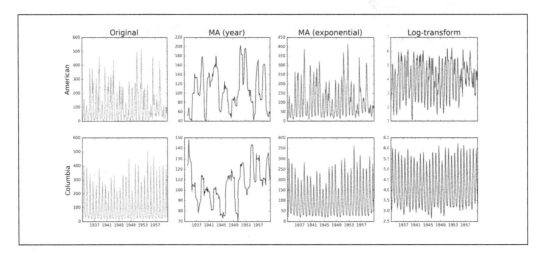

The MA smoothing technique removes much of the noise and reveals the local trends that exist in the data; for the American River, the flow was initially rising up until roughly 1942, falling until 1950, to rise and fall again afterward. For the Columbia River, one can clearly observe an initial declining trend that reverses around 1945.

The exponential smoothing is not as crude as MA and removes only the biggest spikes in the data while still maintaining its overall shape. The log transform normalizes the amplitude present in the data. As mentioned earlier, this is the only technique that is fully reversible and does not cause any data loss.

There's more...

Holt's transform is another example of an exponential smoothing. The difference is that it does not exponentiate the smoothing parameter (the ts_smoothing_alternative.py file):

```python
import numpy as np

def holt_transform(column, alpha):
    '''
        Method to apply Holt transform

        The transform is given as
        y(t) = alpha * x(t) + (1-alpha) y(t-1)
    '''
```

```
# create an np.array from the column
original = np.array(column)

# starting point for the transformation
transformed = [original[0]]

# apply the transform to the rest of the data
for i in range(1, len(original)):
    transformed.append(
        original[i] * alpha +
        (1-alpha) * transformed[-1])

    return transformed
```

The smoothed value of the observation at time t depends on the observed value at time t and preceding smoothed value; the influence of each observation is controlled via the `alpha` parameter.

We first create a NumPy array. The starting time is transformed is the original observation. We then loop through all the observations in the array and apply the logic.

We use Holt's smoothing method with `alpha` set to `0.5`:

```
ma_transformHolt = riverFlows.apply(
    lambda col: holt_transform(col, 0.5), axis=0)
```

The `.apply(...)` method with axis set to `0` passes columns one by one to our `holt_transform(...)` method.

We can also apply differencing that quickly removes the trend from our data and makes it stationary; differencing is a method that calculates a difference between an observation at time t and its precedent (at time $t-1$):

```
difference = riverFlows - riverFlows.shift(-1)
```

Here, we take advantage of the `.shift(...)` method that we explained earlier in this chapter in the *Handling date objects in Python* recipe.

The resulting transformations look as follows:

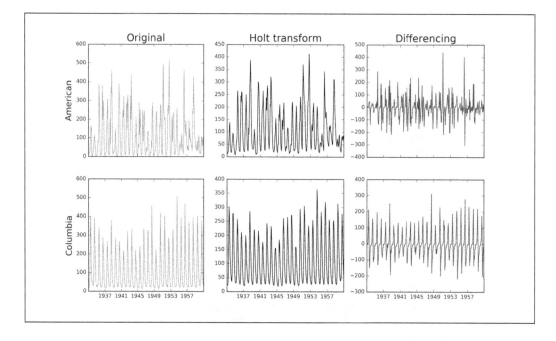

As you can see, the method, just like exponential smoothing, also removes most of the noise but preserves the shape of the data. The differencing is useful if you are more interested in predicting changes in the time series than the value itself; an application of this could be in predicting stock changes.

Filtering the time series data

Removing the noise via smoothing is only one of the techniques. In this recipe, we will see how to use convolution and other filters to extract only certain frequencies from our data.

Getting ready

To execute this recipe, you will need pandas, Statsmodels, NumPy, Scipy, and Matplotlib. No other prerequisites are required.

How to do it...

Convolution, in layman terms, can be understood as an overlap between a function f (our time series) and some function g (our filter). Convolution blurs the time series (and in this sense can be understood as a smoothing technique).

 A good introduction to convolution can be found at http://www.songho.ca/dsp/convolution/convolution.html.

The following script can be found in td_filtering.py:

```
# prepare different filters
MA_filter     = [1] * 12
linear_filter = [d * (1/12) for d in range(0,13)]
gaussian      = sc.signal.gaussian(12, 2)

# convolve
conv_ma       = riverFlows.apply(
    lambda col: sm.tsa.filters.convolution_filter(
        col, MA_filter), axis=0).dropna()

conv_linear   = riverFlows.apply(
    lambda col: sm.tsa.filters.convolution_filter(
        col, linear_filter), axis=0).dropna()

conv_gauss    = riverFlows.apply(
    lambda col: sm.tsa.filters.convolution_filter(
        col, gaussian), axis=0).dropna()
```

How it works...

As the first filter, we are just showing that a moving average filter can be achieved via convolution: the filter is a list of 12 elements, each element is 1.

The second filter linearly increases over the course of 12 periods; the filter gradually decreases the importance of older observations in the output value.

The last filter uses the Gaussian function to filter the data.

The filters' responses look as follows:

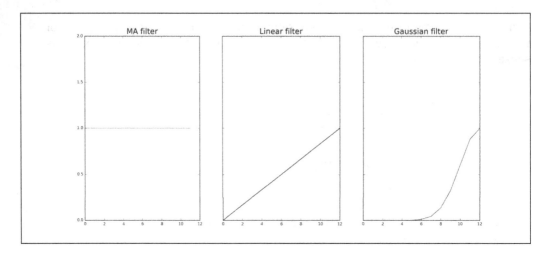

We then use the .convolution_filter(...) method of Statsmodels to filter our dataset. The method accepts the data as its first parameter and the filter array.

Inherently, the methods cannot produce any results for certain observations so we use the .dropna() method to remove the missing observations. The filtered datasets look as follows:

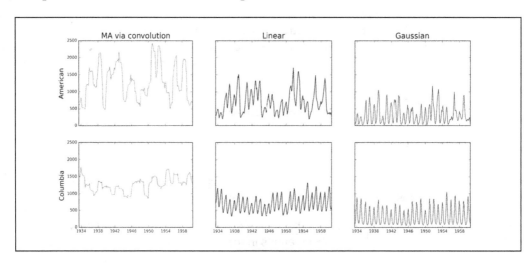

The MA via convolution produces exactly the same results as the .rolling_mean(...) method that we used earlier. The linear filter removes some of the high amplitude spikes from the dataset. Finally, the Gaussian blurring does what it should—it not only reduces the amplitude of the observations, but also makes it smoother.

There's more...

For economic (or natural, in our case) data, some filters are better suited than others: these include **Baxter-King** (**BK**), **Hodrick-Prescott** (**HP**), and **Christiano-Fitzgerald** (**CF**) filters.

 The related scientific papers, for those interested, can be found at http://www.nber.org/papers/w5022.pdf (Baxter-King), https://www0.gsb.columbia.edu/faculty/rhodrick/prescott-hodrick1997.pdf (Hodrick-Prescott), and http://www.nber.org/papers/w7257.pdf (Christiano-Fitzgerald).

We use the Statsmodels implementation of these filters:

```
bkfilter = sm.tsa.filters.bkfilter(riverFlows, 18, 96, 12)
hpcycle, hptrend = sm.tsa.filters.hpfilter(riverFlows, 129600)
cfcycle, cftrend = sm.tsa.filters.cffilter(riverFlows,
    4, 9, False)
```

The BK filter is a bandpass filter; it removes both high and low frequencies from time series. The .bkfilter(...) method takes the data as its first parameter. The next parameter is the minimum period of oscillations; for monthly data, it is suggested to use 18, for quarterly—6, and for yearly—1.5. The next parameter defines the maximum period of oscillations: for monthly data, it is suggested to use 96. The last parameter determines a lead-lag (or a window) that the filter uses.

The HP filter separates the original time series into trend and business cycle components by solving a minimization problem. The .hpfilter(...) takes the data as its first argument and the lamb smoothing parameter. Commonly, a frequency of 1,600 is used for quarterly separated data, 6.25 for annual data, and (as in our case) 129600 for monthly data.

The CF filter is similar to the HP filter in the sense that it decomposes the original time series into trend values and the business cycle component. The .cffilter(...) method, as all the other ones, takes the data as its first parameter. Similarly to the BK filter, the next two parameters specify the minimum and maximum periods of oscillations, respectively. The last parameter, drift, determines whether the trend should be removed from the data.

The filtered data looks as follows:

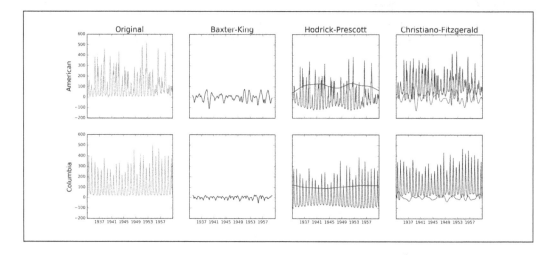

The BF filter removes most of the amplitude from the data and makes it stationary. Analyzing the output of the HP filter, you can clearly see that the Columbia River has almost constant long-term trend while the trend changes over time for the American River. Also, the cyclical components for the American River reveal a similar pattern. The CF filter outputs confirm this: the trend component of the American River is much more volatile than the one of the Columbia River.

Removing trend and seasonality

As mentioned earlier, a time series is stationary if, and only if, its mean does not depend on time, its variance is constant and also does not depend on time, and the autocorrelation does not vary either. This means that any time process with a trend and seasonality is not stationary.

The ARMA and ARIMA models that we will introduce in the next recipe require the data to be stationary (or close to). Thus, in this recipe, you will learn how to remove trend and seasonality from our river flow data.

Getting ready

To execute this recipe, you will need `pandas`, `NumPy`, `Statsmodels`, and `Matplotlib`. No other prerequisites are required.

How to do it...

`Statsmodels` provides convenience methods that will help us detrend and remove the seasonality from our data (the `ts_detrendAndRemoveSeasonality.py` file):

```python
def period_mean(data, freq):
    '''
        Method to calculate mean for each frequency
    '''
    return np.array(
        [np.mean(data[i::freq]) for i in range(freq)])

# read the data
riverFlows = pd.read_csv(data_folder + 'combined_flow.csv',
    index_col=0, parse_dates=[0])

# detrend the data
detrended = sm.tsa.tsatools.detrend(riverFlows,
    order=1, axis=0)

# create a data frame with the detrended data
detrended = pd.DataFrame(detrended, index=riverFlows.index,
    columns=['american_flow_d', 'columbia_flow_d'])

# join to the main dataset
riverFlows = riverFlows.join(detrended)

# calculate trend
riverFlows['american_flow_t'] = riverFlows['american_flow'] \
    - riverFlows['american_flow_d']
riverFlows['columbia_flow_t'] = riverFlows['columbia_flow'] \
    - riverFlows['columbia_flow_d']

# number of observations and frequency of seasonal component
nobs = len(riverFlows)
freq = 12   # yearly seasonality

# remove the seasonality
for col in ['american_flow_d', 'columbia_flow_d']:
    period_averages = period_mean(riverFlows[col], freq)
    riverFlows[col[:-2]+'_s'] = np.tile(period_averages,
        nobs // freq + 1)[:nobs]
    riverFlows[col[:-2]+'_r'] = np.array(riverFlows[col]) \
        - np.array(riverFlows[col[:-2]+'_s'])
```

How it works...

First, as always, we read in the data.

Using the .detrend(...) method of Statsmodels, we then remove the trend from our data. The order parameter specifies the type of the trend: 0 indicates a constant trend, 1 means it is linear, and 2 would attempt to remove a quadratic trend.

The .detrend(...) method returns a NumPy array; in our case, it is a two-dimensional array as we have two columns in the original dataset. Thus, for the ease of joining with our original dataset, in the next step, we create a DataFrame with detrended observations. We are reusing the index from our original DataFrame: riverFlows. Also, in order to avoid collision with the original dataset, we suffix the column names with _d.

The .join(...) method of pandas merges the two DataFrames using their indices (by default): it matches the index values from both DataFrames and returns the corresponding values. The .join(...) method, however, gives you control and allows you to specify the column you want your join to be on; the key-column needs to be present in both of the datasets for it to work. It also allows you to specify how you want to join the DataFrames: by default, it is a left join that returns all the records from the caller DataFrame and all the values from the passed DataFrame that match the key; as we have exactly the same indices in both DataFrames, we can just call the method.

To understand other types of joins, see http://www.sql-join.com. I also recommend perusing the documentation for the .join(...) method at http://pandas.pydata.org/pandas-docs/stable/generated/pandas.DataFrame.join.html.

The detrended time series does not differ much from the original ones:

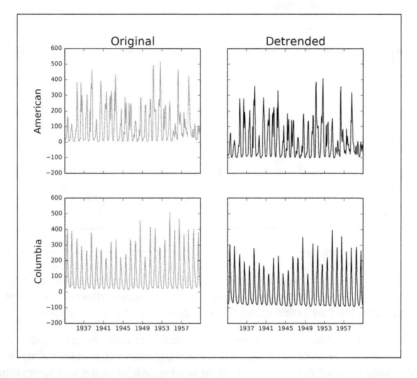

The trend itself is nothing more than the difference between the original values and detrended ones.

Once we have detrended our time series, we can calculate the seasonal component. We expect to have a yearly seasonality so our freq is set to 12. What this means is that we expect that a pattern would repeat every 12 months. The period_mean(...) method calculates exactly this. The data[i::freq] dynamically creates a list of observations by selecting the *i*th element from the list and then every *12*th (freq) following the *i*th.

The subset syntax d[f:l:freq] specifies the first position f, the last position l, and the frequency of sampling freq. Assuming that d = [1,2,3,4,5,6,7,8,9,10] and d[1::2], we would get [2,4,6,8,10].

The `.mean(...)` method of `NumPy` calculates the average of all the elements in such an array. For each river, we then obtain the following:

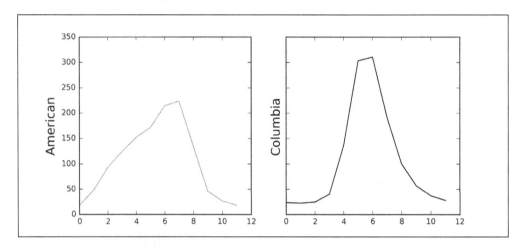

The American River seems to be more tamed: it gradually rises to its peak around August and then falls off toward the end of the year. In contrast, the Columbia River is fairly quiet for most of the year but then the flow jumps up significantly during the summer months.

With the seasonal averages calculated, we can then subtract these from the detrended data. We use the `.tile(...)` method of `NumPy` to repeat the seasonal pattern for the length of our time series. The method takes the pattern to be repeated as the first parameter. The second parameter specifies how many times the pattern should be repeated.

[The `//` operator returns a value of the division rounded down to the nearest integer.]

Finally, we calculate the residuals after removing the seasonality:

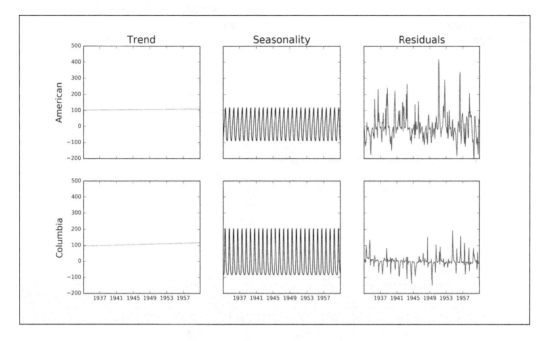

So, here it is: our time series decomposed into a linear trend, seasonal component, and remaining residuals. You can see that the flow for both of the rivers increases over time but the increase is negligible. We can clearly see the yearly seasonal patterns (as expected) with (as previously observed) a higher amplitude of seasonal variations for the Columbia River. As for the residuals, the amplitude of changes of the American River is significantly higher than that of the Columbia River.

There's more...

Statsmodels has another useful method to decompose the time series:.seasonal_decompose(...).

> Sadly, the documentation of Statsmodels is lacking; the project only mentions the .seasonal_decompose(...) method in the release notes with one simple example and the documentation for the method is nowhere to be found. Studying the method's source code on GitHub reveals much more and I encourage you to do so as well (https://github.com/statsmodels/statsmodels/blob/master/statsmodels/tsa/seasonal.py).

The method accepts the data to be decomposed as its first parameter. You can specify the type of the underlying model: it can be either additive (in our case) or multiplicative (by passing m). You also specify the freq parameter; this one might be left out and the frequency will be inferred from the data.

 For a good read on the differences between the additive and multiplicative decomposition models, visit `https://onlinecourses.science.psu.edu/stat510/node/69`.

The method returns a DecomposeResult object. To access the components of the decomposition, you call the attributes of the object:

```
for col in riverFlows.columns:
    # seasonal decomposition of the data
    sd = sm.tsa.seasonal_decompose(riverFlows[col],
        model='a', freq=12)

    riverFlows[col + '_resid'] = sd.resid \
        .fillna(np.mean(sd.resid))

    riverFlows[col + '_trend'] = sd.trend \
        .fillna(np.mean(sd.trend))

    riverFlows[col + '_seas'] = sd.seasonal \
        .fillna(np.mean(sd.seasonal))
```

The resulting decomposition looks different to the one we produced previously:

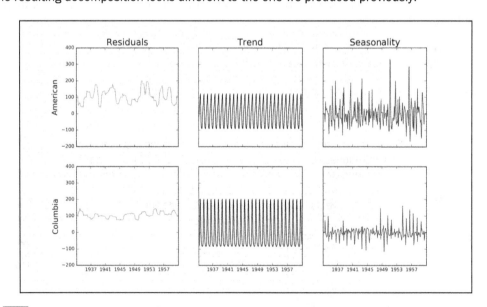

The differences come from a differing way of removing trend: we assumed that the trend is linear in the whole time domain while the `.seasonal_decompose(...)` method uses convolution filtering to uncover the underlying trend. If you look closely, you might start seeing similarities between the preceding trend chart and MA chart in the Filter the time series data recipe; the difference between the trend chart from this method and convolution version of MA is in the weights used by the `.seasonal_decompose(...)` method.

Forecasting the future with ARMA and ARIMA models

The **autoregressive moving average** (**ARMA**) model and its generalization—the **autoregressive integrated moving average** (**ARIMA**) model—are the two most commonly used models to predict the future from time series data. The generalization of the ARIMA model comes from the integrated part: the first step of the model is to differentiate data before estimating the AR and MA parts.

Getting ready

To execute this recipe, you will need `pandas`, `NumPy`, `Statsmodels`, and `Matplotlib`. You will also need the data prepared in the previous recipe. No other prerequisites are required.

How to do it...

We wrap our process within methods so that most of the modeling is automated (the `ts_arima.py` file):

```
def plot_functions(data, name):
    '''
        Method to plot the ACF and PACF functions
    '''
    # create the figure
    fig, ax = plt.subplots(2)

    # plot the functions
    sm.graphics.tsa.plot_acf(data, lags=18, ax=ax[0])
    sm.graphics.tsa.plot_pacf(data, lags=18, ax=ax[1])

    # set titles for charts
    ax[0].set_title(name.split('_')[-1])
    ax[1].set_title('')

    # set titles for rows
    ax[0].set_ylabel('ACF')
```

```
        ax[1].set_ylabel('PACF')

        # save the figure
        plt.savefig(data_folder+'/charts/'+name+'.png',
            dpi=300)

def fit_model(data, params, modelType, f, t):
    '''
        Wrapper method to fit and plot the model
    '''
    # create the model object
    model = sm.tsa.ARIMA(data, params)

    # fit the model
    model_res = model.fit(maxiter=300, trend='nc',
        start_params=[.1] * (params[0]+params[2]), tol=1e-06)

    # plot the model
    plot_model(data['1950':], model_res, params,
        modelType, f, t)

    # and save the figure
    plt.savefig(data_folder+'/charts/'+modelType+'.png',
        dpi=300)

def plot_model(data, model, params, modelType, f, t):
    '''
        Method to plot the predictions of the model
    '''
    # create figure
    fig, ax = plt.subplots(1, figsize=(12, 8))

    # plot the data
    data.plot(ax=ax, color=colors[0])

    # plot the forecast
    model.plot_predict(f, t, ax=ax, plot_insample=False)

    # define chart text
    chartText = '{0}: ({1}, {2}, {3})'.format(
        modelType.split('_')[0], params[0],
        params[1], params[2])

    # and put it on the chart
    ax.text(0.1, 0.95, chartText, verticalalignment='top',
        horizontalalignment='left',transform=ax.transAxes)
```

How it works...

After reading the data in, we first look at the ACF and PACF functions of the residuals:

```
plot_functions(riverFlows['american_flow_r'],
    'ACF_PACF_American')
plot_functions(riverFlows['columbia_flow_r'],
    'ACF_PACF_Columbia')
```

Analyzing the charts will help us determine the AR and MA components for our models:

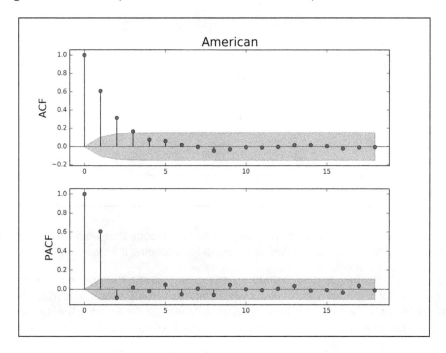

The preceding and following graphs represent the ACF and PACF function of America and Columbia respectively:

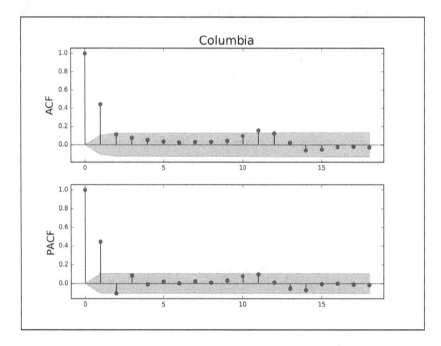

The `plot_functions(...)` method is very similar to the code that we developed in the *Understanding time series data* recipe so we will skip describing it here.

Looking at the charts, we can read the autoregressive and moving average components that we can use as the starting point of our model building efforts: the ACF function helps determine the order of the MA while the PACF function allows us to determine the AR part. The rule of thumb is to look at the last observation that lies significantly outside of the confidence interval.

From our charts, we can deduce the following: we will start with AR(2) and MA(4) for the American River model and AR(3) and MA(2) for the Columbia River model:

```
# fit american models
fit_model(riverFlows['american_flow_r'], (2, 0, 4),
    'ARMA_American', '1960-11-30', '1962')
fit_model(riverFlows['american_flow_r'], (2, 1, 4),
    'ARIMA_American', '1960-11-30', '1962')

# fit colum models
fit_model(riverFlows['columbia_flow_r'], (3, 0, 2),
    'ARMA_Columbia', '1960-09-30', '1962')
fit_model(riverFlows['columbia_flow_r'], (3, 1, 2),
    'ARIMA_Columbia', '1960-09-30', '1962')
```

The `fit_model(...)` method encapsulates all the necessary steps of our model building. It takes the data to be used to estimate the model as its first parameter. The `params` argument is used to define the model parameters. The `modelType` parameter is used to decorate the chart when we call the `plot_model(...)` method; the `f` (from) and `t` (to) parameters are also passed to `plot_model(...)`.

First, we create the model object. We chose to use the `.ARIMA(...)` model only as it can be specialized to become the ARMA model. The `.ARIMA(...)` method takes the data as its first parameter and the tuple of params as the second parameter. The first element of the tuple describes the AR part of the model, the second one describes the lag to be used for the differencing (in the ARIMA model), and the last element defines the MA component.

 By setting the differencing part to 0, the ARIMA model becomes the ARMA model.

Next, we fit the model. We set the maximum number of iterations to 300 and trend with no constant. We also define `start_params`; we start with a list of six elements, the sum of all the AR and MA components. Each element of our list is set to a starting point of `0.1`. The tolerance is set to be at `0.000001`; if the difference in error reduction between iterations is lower than the tolerance, the estimation stops.

Once the model is estimated, we will plot how it performs. The `plot_model(...)` method starts with creating the plot and plotting the observed data. We limit the data to start from 1950 (for brevity of the chart) when we call the method. The `.plot_predict(...)` method uses the estimated model's parameters to predict the observations for the future time. The first two parameters specify from and to boundaries of the prediction: we can choose the starting and ending points. We also set `plot_insample` to `False`; while the documentation on this parameter is highly insufficient, we empirically determined that this parameter prevents the method from plotting the observations that overlap with the predictions in a different color.

 You should not be reaching too far into the future as the confidence interval for such a prediction would widen beyond any reason.

We also put a text in the top left corner of the chart; we use the `.text(...)` method to do this. The first two parameters determine the coordinates on the chart: by transforming the Axes coordinates with `.transformAxes`, we know that the coordinate `(0, 0)` is in the bottom left corner and point `(1, 1)` is in the top right one. The third parameter is the text that we want to place on the chart.

Now that we have the model estimated, let's look at the charts:

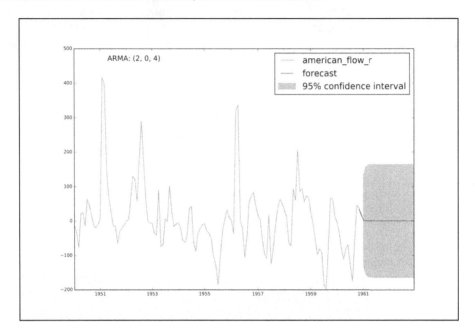

The predicted flow of the American river initially follows a short-term trend but later on (as we are moving further away from the last observation), the confidence interval explodes and the prediction becomes a constant:

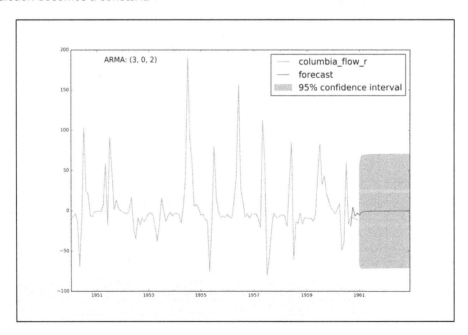

For the Columbia River, the initial predicted river flow seems to diverge from the observed values but flattens quickly:

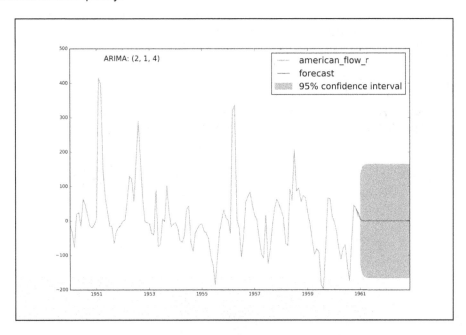

The predictions of the ARIMA model did not differ significantly from the ARMA model; the prediction seems to initially follow the observed data in order to flatten later on:

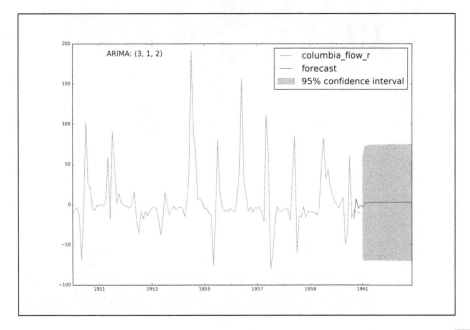

The models predict what essentially can be thought of as an average of the residuals.

 Check John Wittenauer's efforts at predicting the S&P500 index to see that this is not really unexpected and that predicting time series is not a trivial matter (`http://www.johnwittenauer.net/a-simple-time-series-analysis-of-the-sp-500-index/`).

The AR and MA parameters that we got from the ACF and PACF functions should be used just as a starting point. If the models perform well, keep them. Otherwise, you can change them slightly as long as you do not depart far from these initial values.

We should also look at the distribution of the error terms of our models:

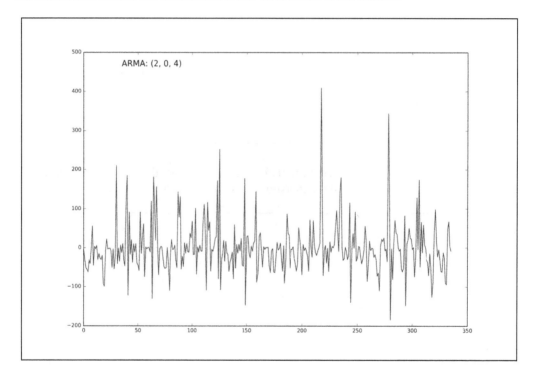

The error terms of the ARMA model for the American River seem to be fairly random and this is something that anyone predicting time series should be aiming at:

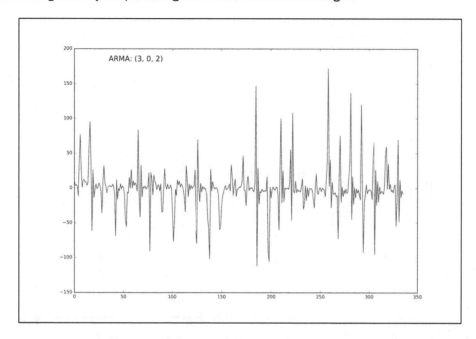

As for the Columbia River, some recurring patterns can be spotted in the error terms, eventually leading to questioning the ability of the model to predict future water flows:

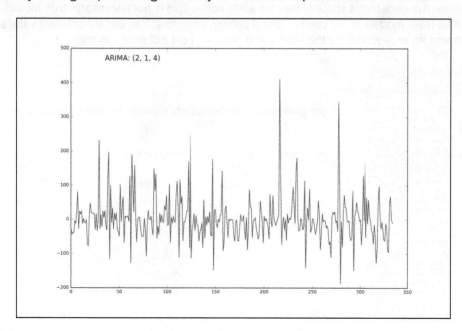

The ARIMA model for the American River also produces what looks like random residuals:

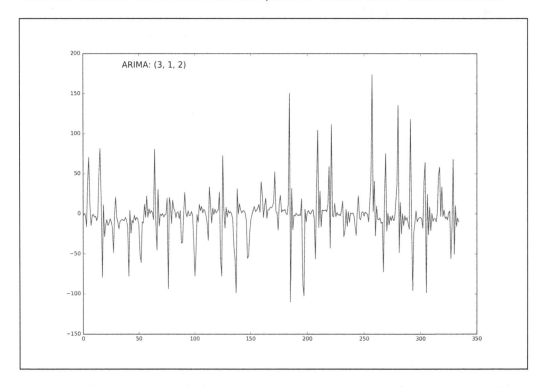

Ultimately, the error terms should look like white noise. This is not necessarily true for the Columbia River models as we can still see a pattern emerging. You can try different values of the AR and MA parameters for the model and see if you get different results.

See also

McKinney, Perktold, and Seabold gave a nice presentation/tutorial on time series. It's worth going through.

http://conference.scipy.org/scipy2011/slides/mckinney_time_series.pdf

8

Graphs

In this chapter, you will learn the following recipes:

- ▶ Handling graph objects in Python with NetworkX
- ▶ Using Gephi to visualize graphs
- ▶ Identifying people whose credit card details were stolen
- ▶ Identifying those responsible for stealing the credit cards

Introduction

Graphs are everywhere; when you get in your car and drive around using a GPS, you perhaps do not even realize that it is solving a graph problem to get you from point A to point B over the shortest path or in the shortest time.

The origins of graph theory reach the 18th century when Leonard Euler proposed the solution to the Königsberg bridge problem. (To read more on the topic, you can refer to `http://www2.gsu.edu/~matgtc/origin%20of%20graph%20theory.pdf`.) From that point onward, some of the problems that were deemed unsolvable could be solved; the Internet (or even your local network) can be viewed and analyzed as a graph, scheduling problems that airlines solve can be modeled as a graph, or (as we will see) a social network is much easier to handle if we realize that it is a graph.

Graphs are structures consisting of nodes (sometimes called vertices) and edges (sometimes called arcs or lines) that connect two nodes:

The preceding example shows the simplest network possible, with two nodes and one edge connecting them. This example shows an undirected graph: a graph where the connection between the two nodes has no direction. For example, if nodes were people, the edge might be representing the fact that they know each other. In a situation like this, we cannot possibly identify a direction as the relationship is bi-directional: A knows B and B knows A.

This might not be the case if there is an implied direction of progression through the graph. Such an example can be an engineering plan of a building or bridge where nodes are the specific tasks that the contractors have to do and edges represent the progression through such a plan.

Handling graph objects in Python with NetworkX

The boom of social networks such as Facebook, Twitter, or LinkedIn (among others) has introduced a wide set of problems. Examples of these (but not limited to) are: who is a friend of whom, can I reach a recruiter in that particular company that I would love to work for via my network of friends and associates, am I connected to President Obama in any way, and who is the most influential person in my network?

These types of problems are very common in the present society and any data scientist should know how to tackle them.

In this recipe, we will introduce a fabricated Twitter network of 20 people. You will learn how to create a graph, add nodes and edges (and additional metadata), analyze the graph, and export it so that we can read it using Gephi.

Getting ready

To execute this recipe, you will need `NetworkX`, `collections`, and `Matplotlib`. No other prerequisites are required.

How to do it...

`NetworkX` provides you with a framework to build and analyze graphs. The package can handle undirected as well as directed graphs. What is more, it is also possible to model multi-graphs: graphs with more than one edge between the same two nodes and graphs with self-loops.

If you are using Anaconda, chances are that `NetworkX` is already installed on your machine. To check, run this command:

```
conda list | grep networkx
```

If the results indicate that `NetworkX` is on your system, you are good to proceed. If, however, the package cannot be found, you need to install it. Issue the following command to do so:

```
conda install networkx
```

You should see something like the following screenshot:

```
Fetching package metadata: ....
Solving package specifications: ..........

Package plan for installation in environment /Users/drabast/anaconda:

The following packages will be downloaded:

    package                    |           build
    ---------------------------|-----------------
    conda-4.0.5                |          py35_0       188 KB
    networkx-1.11              |          py35_0       1.1 MB
    ---------------------------------------------------------
                                           Total:       1.3 MB

The following NEW packages will be INSTALLED:

    networkx: 1.11-py35_0

The following packages will be UPDATED:

    conda:    4.0.4-py35_0 --> 4.0.5-py35_0

Proceed ([y]/n)? y

Fetching packages ...
conda-4.0.5-py 100% |####################| Time: 0:00:00 400.26 kB/s
networkx-1.11- 100% |####################| Time: 0:00:02 440.62 kB/s
Extracting packages ...
[      COMPLETE      ]|#########################################| 100%
Unlinking packages ...
[      COMPLETE      ]|#########################################| 100%
Linking packages ...
[      COMPLETE      ]|#########################################| 100%
```

We have just installed version 1.11 of `NetworkX`. You can check this by executing the following command:

```
python -c "import networkx; print(networkx.__version__)"
```

This will produce the following output:

```
1.11
```

If you, however, do not run Anaconda, you can go to `https://github.com/networkx/networkx/`, download the whole source code, unzip it, and issue the following commands (assuming that you unzipped the code to the `networkx` folder):

cd networkx

python setup.py install

You're now ready to proceed. The script that we will use in this recipe is located in the `graph_handling.py` file in the folder for this chapter (`Codes/Chapter08`):

```python
import networkx as nx
import networkx.algorithms as alg
import numpy as np
import matplotlib.pyplot as plt

# create graph object
twitter = nx.Graph()
```

How it works...

First, we import the necessary modules; `networkx.algorithms` gives us access to a substantial set of graph algorithms that we will make use of later.

Next, we create a skeleton of our undirected graph with a single edge between nodes.

> For a list of all the graph types, check `NetworkX`'s documentation:
> `http://networkx.github.io/documentation/networkx-1.10/reference/classes.html`

Now, let's add some nodes. As mentioned earlier, we have 20 people in our network. We add them using the `.add_node(...)` method:

```python
# add users
twitter.add_node('Tom', {'age': 34})
twitter.add_node('Rachel', {'age': 33})
twitter.add_node('Skye', {'age': 29})
twitter.add_node('Bob', {'age': 45})
twitter.add_node('Mike', {'age': 23})
twitter.add_node('Peter', {'age': 46})
twitter.add_node('Matt', {'age': 58})
twitter.add_node('Lester', {'age': 65})
twitter.add_node('Jack', {'age': 32})
twitter.add_node('Max', {'age': 75})
twitter.add_node('Linda', {'age': 23})
twitter.add_node('Rory', {'age': 18})
```

```
twitter.add_node('Richard', {'age': 24})
twitter.add_node('Jackie', {'age': 25})
twitter.add_node('Alex', {'age': 24})
twitter.add_node('Bart', {'age': 33})
twitter.add_node('Greg', {'age': 45})
twitter.add_node('Rob', {'age': 19})
twitter.add_node('Markus', {'age': 21})
twitter.add_node('Glenn', {'age': 24})
```

The method accepts a node ID as its first argument. The second argument is optional; it is a dictionary of metadata that further refines the node.

> Nodes need to be distinct, that is, if you had two Peters, you would have to be able to distinguish them somehow (using their last name or sequential number is fine).
>
> You can also add nodes from a list. Check the .add_nodes_ from(...) method at http://networkx.github.io/ documentation/networkx-1.10/reference/generated/ networkx.Graph.add_nodes_from.html#networkx. Graph.add_nodes_from.

The metadata can also be added by accessing the node directly. As this should be a graph resembling the Twitter social network, let's add the number of posts that each user tweeted:

```
# add posts
twitter.node['Rory']['posts'] = 182
twitter.node['Rob']['posts'] = 111
twitter.node['Markus']['posts'] = 159
twitter.node['Linda']['posts'] = 128
twitter.node['Mike']['posts'] = 289
twitter.node['Alex']['posts'] = 188
twitter.node['Glenn']['posts'] = 252
twitter.node['Richard']['posts'] = 106
twitter.node['Jackie']['posts'] = 138
twitter.node['Skye']['posts'] = 78
twitter.node['Jack']['posts'] = 62
twitter.node['Bart']['posts'] = 38
twitter.node['Rachel']['posts'] = 89
twitter.node['Tom']['posts'] = 23
twitter.node['Bob']['posts'] = 21
twitter.node['Greg']['posts'] = 41
twitter.node['Peter']['posts'] = 64
twitter.node['Matt']['posts'] = 8
twitter.node['Lester']['posts'] = 4
twitter.node['Max']['posts'] = 2
```

As you can see, you can call the node by its ID and set the metadata that way; it is useful if you want to set some fixed parameters when you create the node (such as the age that we specified earlier) and then keep updating the variable metadata as things change, for example, when the user submits more posts.

Now, let's see who knows who:

```
# add followers
twitter.add_edge('Rob', 'Rory', {'Weight': 1})
twitter.add_edge('Markus', 'Rory', {'Weight': 1})
twitter.add_edge('Markus', 'Rob', {'Weight': 5})
twitter.add_edge('Mike', 'Rory', {'Weight': 1})
twitter.add_edge('Mike', 'Rob', {'Weight': 1})
twitter.add_edge('Mike', 'Markus', {'Weight': 1})
twitter.add_edge('Mike', 'Linda', {'Weight': 5})
twitter.add_edge('Alex', 'Rob', {'Weight': 1})
twitter.add_edge('Alex', 'Markus', {'Weight': 1})
twitter.add_edge('Alex', 'Mike', {'Weight': 1})
twitter.add_edge('Glenn', 'Rory', {'Weight': 1})
twitter.add_edge('Glenn', 'Rob', {'Weight': 1})
twitter.add_edge('Glenn', 'Markus', {'Weight': 1})
twitter.add_edge('Glenn', 'Linda', {'Weight': 2})
twitter.add_edge('Glenn', 'Mike', {'Weight': 1})
twitter.add_edge('Glenn', 'Alex', {'Weight': 1})
twitter.add_edge('Richard', 'Rob', {'Weight': 1})
twitter.add_edge('Richard', 'Linda', {'Weight': 1})
twitter.add_edge('Richard', 'Mike', {'Weight': 1})
twitter.add_edge('Richard', 'Alex', {'Weight': 1})
twitter.add_edge('Richard', 'Glenn', {'Weight': 1})
twitter.add_edge('Jackie', 'Linda', {'Weight': 1})
twitter.add_edge('Jackie', 'Mike', {'Weight': 1})
twitter.add_edge('Jackie', 'Glenn', {'Weight': 1})
twitter.add_edge('Jackie', 'Skye', {'Weight': 1})
twitter.add_edge('Tom', 'Rachel', {'Weight': 5})
twitter.add_edge('Rachel', 'Bart', {'Weight': 1})
twitter.add_edge('Tom', 'Bart', {'Weight': 2})
twitter.add_edge('Jack', 'Skye', {'Weight': 1})
twitter.add_edge('Bart', 'Skye', {'Weight': 1})
twitter.add_edge('Rachel', 'Skye', {'Weight': 1})
twitter.add_edge('Greg', 'Bob', {'Weight': 1})
twitter.add_edge('Peter', 'Greg', {'Weight': 1})
twitter.add_edge('Lester', 'Matt', {'Weight': 1})
twitter.add_edge('Max', 'Matt', {'Weight': 1})
twitter.add_edge('Rachel', 'Linda', {'Weight': 1})
twitter.add_edge('Tom', 'Linda', {'Weight': 1})
```

```
twitter.add_edge('Bart', 'Greg', {'Weight': 2})
twitter.add_edge('Tom', 'Greg', {'Weight': 2})
twitter.add_edge('Peter', 'Lester', {'Weight': 2})
twitter.add_edge('Tom', 'Mike', {'Weight': 1})
twitter.add_edge('Rachel', 'Mike', {'Weight': 1})
twitter.add_edge('Rachel', 'Glenn', {'Weight': 1})
twitter.add_edge('Lester', 'Max', {'Weight': 1})
twitter.add_edge('Matt', 'Peter', {'Weight': 1})
```

The `.add_edge(...)` method takes the origin node as its first argument and target node as its second argument; in this particular graph, the sequence does not matter as it is an undirected graph. You can only provide these two parameters; the dictionary with metadata is optional (as in the case with nodes). We use the `Weight` parameter to differentiate certain connections (and this will become apparent in the next recipe). Let's add the relationships to illustrate these differences:

```
# add relationship
twitter['Rob']['Rory']['relationship'] = 'friend'
twitter['Markus']['Rory']['relationship'] = 'friend'
twitter['Markus']['Rob']['relationship'] = 'spouse'
twitter['Mike']['Rory']['relationship'] = 'friend'
twitter['Mike']['Rob']['relationship'] = 'friend'
twitter['Mike']['Markus']['relationship'] = 'friend'
twitter['Mike']['Linda']['relationship'] = 'spouse'
twitter['Alex']['Rob']['relationship'] = 'friend'
twitter['Alex']['Markus']['relationship'] = 'friend'
twitter['Alex']['Mike']['relationship'] = 'friend'
twitter['Glenn']['Rory']['relationship'] = 'friend'
twitter['Glenn']['Rob']['relationship'] = 'friend'
twitter['Glenn']['Markus']['relationship'] = 'friend'
twitter['Glenn']['Linda']['relationship'] = 'sibling'
twitter['Glenn']['Mike']['relationship'] = 'friend'
twitter['Glenn']['Alex']['relationship'] = 'friend'
twitter['Richard']['Rob']['relationship'] = 'friend'
twitter['Richard']['Linda']['relationship'] = 'friend'
twitter['Richard']['Mike']['relationship'] = 'friend'
twitter['Richard']['Alex']['relationship'] = 'friend'
twitter['Richard']['Glenn']['relationship'] = 'friend'
twitter['Jackie']['Linda']['relationship'] = 'friend'
twitter['Jackie']['Mike']['relationship'] = 'friend'
twitter['Jackie']['Glenn']['relationship'] = 'friend'
twitter['Jackie']['Skye']['relationship'] = 'friend'
    twitter['Tom']['Rachel']['relationship'] = 'spouse'
twitter['Rachel']['Bart']['relationship'] = 'friend'
twitter['Tom']['Bart']['relationship'] = 'sibling'
```

```
twitter['Jack']['Skye']['relationship'] = 'friend'
twitter['Bart']['Skye']['relationship'] = 'friend'
twitter['Rachel']['Skye']['relationship'] = 'friend'
twitter['Greg']['Bob']['relationship'] = 'friend'
twitter['Peter']['Greg']['relationship'] = 'friend'
twitter['Lester']['Matt']['relationship'] = 'friend'
twitter['Max']['Matt']['relationship'] = 'friend'
twitter['Rachel']['Linda']['relationship'] = 'friend'
twitter['Tom']['Linda']['relationship'] = 'friend'
twitter['Bart']['Greg']['relationship'] = 'sibling'
twitter['Tom']['Greg']['relationship'] = 'sibling'
twitter['Peter']['Lester']['relationship'] = 'generation'
twitter['Tom']['Mike']['relationship'] = 'friend'
twitter['Rachel']['Mike']['relationship'] = 'friend'
twitter['Rachel']['Glenn']['relationship'] = 'friend'
twitter['Lester']['Max']['relationship'] = 'friend'
twitter['Matt']['Peter']['relationship'] = 'friend'
```

We can have four types of relationships in our network: friend, spouse, sibling, and generation. The last one is the parent-child relationship. We gave Weight of 1, 5, 2, and 2 (respectively) to each type of relationship.

Note how we access nodes and set the metadata by simply calling the graph with the edge that we want to access, for example, twitter['Rachel']['Tom'], and then set the attribute of interest.

There's more...

NetworkX provides a vast array of useful methods that help in accessing, manipulating, and analyzing the graphs.

To retrieve a list of all the nodes from the graph, you can call the .nodes(...) method without any parameters, that is, .nodes(). If you print such a list, you will get something as follows:

```
Just nodes:  ['Rob', 'Glenn', 'Linda', 'Rory', 'Alex', 'Matt', 'Mike',
'Skye', 'Greg', 'Rachel', 'Bart', 'Markus', 'Tom', 'Jack', 'Richard',
'Jackie', 'Peter', 'Lester', 'Bob', 'Max']
```

The `.nodes(...)` method also accepts the `data` parameter; the `.nodes(data=True)` call will also return the metadata for each node:

```
Nodes with data:  [('Rob', {'age': 19, 'posts': 111}), ('Glenn',
{'age': 24, 'posts': 252}), ('Linda', {'age': 23, 'posts': 128}),
('Rory', {'age': 18, 'posts': 182}), ('Alex', {'age': 24, 'posts':
188}), ('Matt', {'age': 58, 'posts': 8}), ('Mike', {'age': 23,
'posts': 289}), ('Skye', {'age': 29, 'posts': 78}), ('Greg', {'age':
45, 'posts': 41}), ('Rachel', {'age': 33, 'posts': 89}), ('Bart',
{'age': 33, 'posts': 38}), ('Markus', {'age': 21, 'posts': 159}),
('Tom', {'age': 34, 'posts': 23}), ('Jack', {'age': 32, 'posts':
62}), ('Richard', {'age': 24, 'posts': 106}), ('Jackie', {'age':
25, 'posts': 138}), ('Peter', {'age': 46, 'posts': 64}), ('Lester',
{'age': 65, 'posts': 4}), ('Bob', {'age': 45, 'posts': 21}), ('Max',
{'age': 75, 'posts': 2})]
```

Edges can be accessed in a similar fashion; by calling the `.edges(...)` method; a call to `.edges(Data=True)` will return the following list (abbreviated):

```
Edges with data:  [('Rob', 'Rory', {'relationship': 'friend',
'Weight': 1}), ('Rob', 'Alex', {'relationship': 'friend', 'Weight':
1}), ('Rob', 'Markus', {'relationship': 'spouse', 'Weight': 5}),
('Rob', 'Mike', {'relationship': 'friend', 'Weight': 1}), ('Rob',
'Richard', {'relationship': 'friend', 'Weight': 1}), ('Rob', 'Glenn',
{'relationship': 'friend', 'Weight': 1}), ...]
```

Having created the graph, let's analyze its structure. The first metric that we will be looking at is the graph's density:

```
print('\nDensity of the graph: ', nx.density(twitter))
```

The `.density(...)` parameter measures the degree of connectivity between the nodes in a graph; a graph with all the nodes connected with all the other nodes (with no self-loops) would have a density of 1. Simply put, the density of a graph is the ratio of the number of edges in such a graph to the total number of all possible edges connecting all the nodes in a graph. For our graph, we get the following:

```
Density of the graph:  0.23684210526315788
```

It shows that our graph is sparse: the total possible number of edges in our graph is given by the equation $n * (n-1) / 2$, so we get $20 * 19 / 2 = 190$ possible edges. The total number of edges in our graph is 45. Thus, only 23.7% of all the possible connections are present in our graph.

 If you do not know where the equation $n * (n-1) / 2$ comes from, check the following link:

http://jwilson.coe.uga.edu/EMAT6680Fa2013/
Hendricks/Essay%202/Essay2.html

Another useful metric of a graph is its degree. The degree of a node is the total number of its neighbors. The neighborhood of a node is a list of all the nodes that the node in question is directly connected with via an edge. Thus, the degree of a node is nothing more but a count of the nodes adjacent to the node in question.

The `.centrality.degree_centrality(...)` method calculates a ratio of the degree of the node to the possible maximum degree of a graph (that equals to the number of nodes less 1):

```
centrality = sorted(
    alg.centrality.degree_centrality(twitter).items(),
    key=lambda e: e[1], reverse=True)
```

We sort the results in a descending order given the calculated degree of centrality; in this way, we can see who is the most connected person in our graph:

```
Centrality of nodes:  [('Mike', 0.5263157894736842), ('Glenn',
0.47368421052631576), ('Rachel', 0.3157894736842105), ('Rob',
0.3157894736842105), ('Linda', 0.3157894736842105), ('Alex',
0.2631578947368421), ('Markus', 0.2631578947368421), ('Tom',
0.2631578947368421), ('Richard', 0.2631578947368421), ('Greg',
0.21052631578947367), ('Jackie', 0.21052631578947367), ('Bart',
0.21052631578947367), ('Skye', 0.21052631578947367), ('Rory',
0.21052631578947367), ('Lester', 0.15789473684210525), ('Peter',
0.15789473684210525), ('Matt', 0.15789473684210525), ('Max',
0.10526315789473684), ('Bob', 0.05263157894736842), ('Jack',
0.05263157894736842)]
```

It turns out that Mike and Glenn are the two most connected people in our graph. (In the next recipe we will inspect this visually.)

The `.assortativity.average_neighbor_degree(...)` method, for each node, calculates an average of its neighbors' degrees. It is a metric that allows us to find persons who are friends with the most connected people in our network—a useful metric if you want to get connected with someone you do not know:

```
average_degree = sorted(
    alg.assortativity.average_neighbor_degree(twitter)\
    .items(), key=lambda e: e[1], reverse=True)
```

Let's see who is connected with the most influential people:

```
Average degree:  [('Rory', 7.5), ('Jackie', 7.25), ('Richard',
7.2), ('Alex', 7.0), ('Markus', 6.8), ('Linda', 6.5), ('Rachel',
6.333333333333333), ('Rob', 6.333333333333333), ('Tom', 6.0),
('Glenn', 5.666666666666667), ('Mike', 5.5), ('Bart', 4.75), ('Bob',
4.0), ('Jack', 4.0), ('Skye', 3.75), ('Peter', 3.3333333333333335),
('Greg', 3.25), ('Max', 3.0), ('Matt', 2.6666666666666665), ('Lester',
2.6666666666666665)]
```

Turns out that `Rory`, `Jackie`, `Richard`, and `Alex` are your best bets at expanding your network as these guys know both `Mike` and `Glenn`.

 There's a plethora of other useful metrics that you might want to explore. For starters, you might want to check this website, `http://webwhompers.com/graph-theory. html`, and study the documentation of `NetworkX`, `http://networkx.github.io/documentation/ networkx-1.10/reference/algorithms.html`.

The `NetworkX` framework also has some primitive drawing capabilities built in.

 `NetworkX` can also leverage `Graphviz` and `pydot` but as these two modules work only in Python 2.7 and have not yet been ported to Python 3.4, we were unsuccessful in leveraging their power.

To draw the network that we so meticulously created, we call the `.draw_networkx(...)` method:

```
# draw the graph
nx.draw_networkx(twitter)
plt.savefig('../../Data/Chapter08/twitter_networkx.png')
```

The resulting graph looks somewhat unappealing and not really informative as the content is overlapping and it is really hard to read. Even then, it still shows the structure of our graph. Note that your graph, even though would have exactly the same connections, will most likely have a different layout.

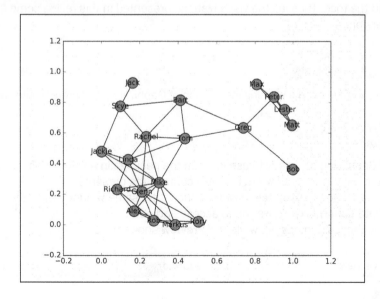

Fortunately, we can export our graph to the GraphML format that can then be digested by Gephi, our next stop at understanding our social network:

```
# save graph
nx.write_graphml(twitter,
    '../../Data/Chapter08/twitter.graphml')
```

See also

See another quick introduction to `NetworkX` at `http://www.python-course.eu/networkx.php`.

Using Gephi to visualize graphs

Gephi is an open source application for the visualization and analysis of complex networks. The package can run on any platform that runs Java, so you can use it in Windows, Linux, or Mac environments.

To obtain your copy of Gephi, go to `https://gephi.org/users/download/` and download the package appropriate for your system. After downloading, follow the prompts to install the package.

We have had many problems trying to run version 0.8.2-beta of Gephi on Mac OS X El Capitan. The newest version of Gephi uses Java libraries that do not work well with the Java version shipped with the Mac OS X (see, for example, `https://github.com/gephi/gephi/issues/1141`). Even installing the legacy package with Java 6 as suggested in the aforementioned thread did not allow us to run Gephi 0.8.2-beta; only downgrading to version 0.8.1-beta did the trick. Thus, all the visualizations presented in this recipe come from the 0.8.1-beta version.

Getting ready

For this recipe, you will need a working installation of Gephi. No other prerequisites are required.

How to do it...

On installing Gephi, you are ready to use it. Open the package in a way specific to your platform. At the top of the window are the view controls (as presented in the view below); the application defaults to the **Overview** view that shows the same windows as in the preceding image. The **Data Laboratory** allows you to access and edit the underlying data that creates the graph (both nodes and edges), while the **Preview** shows the print preview:

The graph controls give you control over how nodes and edges are visualized; these controls allow you to change the colors of the nodes and edges and their size and add labels.

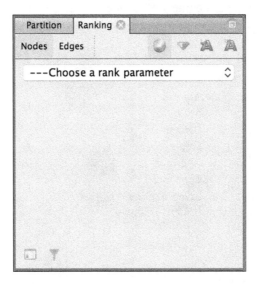

The layout control gives you control over the layout of the graph; we will see how to use the algorithms included shortly.

The statistics and filters section allows you to calculate the statistics of our chart (such as average degree or graph density that we saw in the previous recipe). It also allows you to filter out some of the nodes or edges so that you can focus on just a portion of the graph that is of interest to you.

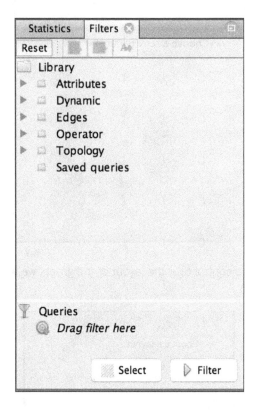

The graph window shows the graph. Go to **File** | **Open**, navigate to the **Data/Chapter08** folder, and select **twitter.graphml**. You should see something similar to the following when the graph opens:

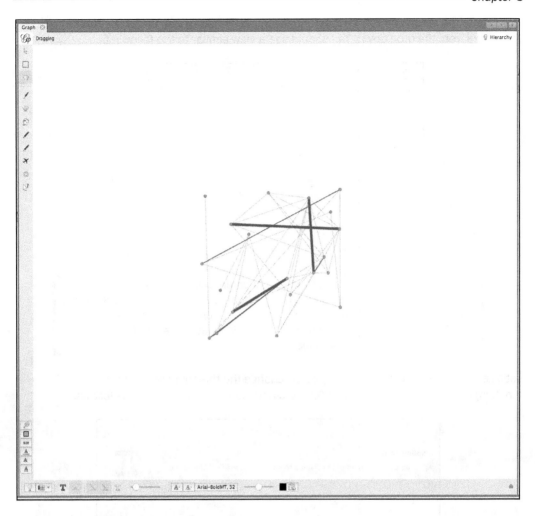

Not really an informative view. The first thing that I usually like to do is color the nodes so that I know the age of the population (we will use a linear scale) and alter the size of the nodes to represent the number of posts:

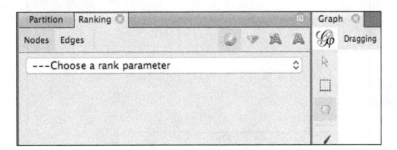

Navigate to the graph controls section and select the **Ranking** tab. Under the **Nodes** tab, select **age** from the drop-down. You should see a view similar to the following:

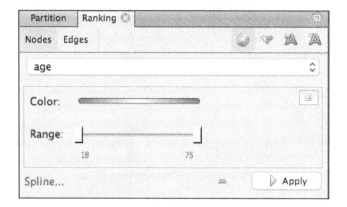

 The new version of Gephi might differ from what you see here and the **Color** option might be showing different choices. For the sake of understanding your ways around Gephi, this should not be a big problem and you should be able to easily follow the examples in this recipe.

Let's keep the colors as they are but slightly change the threshold points (or the transfer function) of the color range. Click on **Spline...** and alter the curve to look as follows:

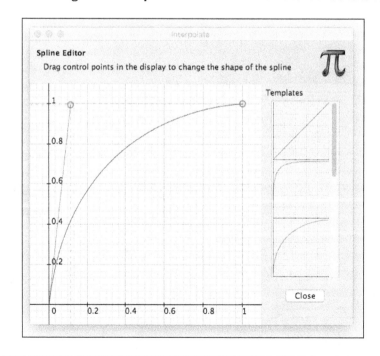

Click on **Apply** and the colors of your nodes should have changed, although you might not really notice it as the node sizes are the same; let's fix this now.

Still under the **Nodes** tab, first select the diamond-looking icon in the top-right corner of the graph controls section; when you hover over the diamond-looking icon, it will read Size/Weight. Now, select **posts** from the drop-down. You should see something similar to this:

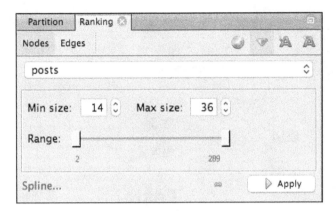

Click on **Apply** and the sizes of the nodes should now be reflective of the number of posts:

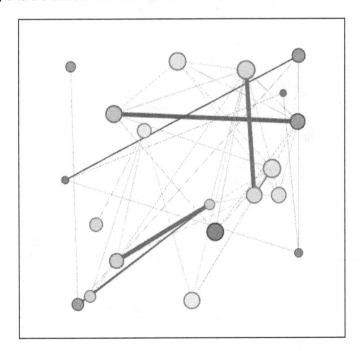

Now the colors can be recognized. However, to understand which node represents whom, let's put some labels on each node.

Navigate to the graph window. At the bottom of the window, you can see a row of icons. Click on the one with a **T** on it (highlighted in the following screenshot):

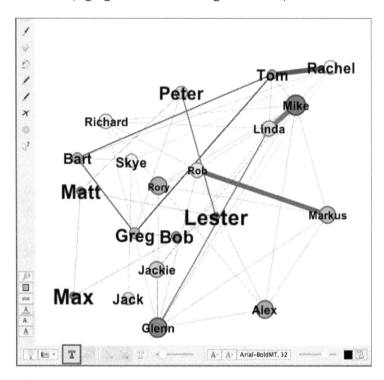

Now we know who is who.

Still, the graph does not reveal any particular shape. So, let's reveal a more informative structure. Navigate to the **Layout** control tab. From the drop-down, choose **Force Atlas**—an algorithm that will help us reveal the hidden form of our graph.

The algorithm balances the tree of the graph in such a way that the connected nodes stick together by means of some gravitational pull while those disconnected tend to repel. You can control the strength of both forces the layout control window (as shown in the following figure):

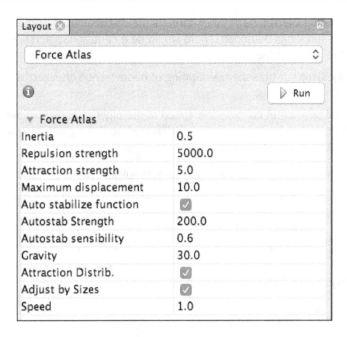

The **Force Atlas** algorithm iteratively arrives at the best layout given the specified parameters:

- The **Inertia** parameter controls how much speed a node retains in each pass; a value of **0.5** means that the nodes will be mostly stationary.

- The **Repulsion strength** parameter defines the strength of the repelling force; a value of **5000.0** means that the graph will be fairly dispersed. Compare this with a value of 15,000 image below on the right.

- **Attraction strength** defines the gravitational pull between connected nodes; the difference between the repulsion and attraction is that the repulsion applies to all the nodes while the gravitational pull works only for the nodes that are connected.

- **Maximum displacement** limits the maximum distance that the node can travel from its original position.

- The **Auto stabilize function**, as the name suggests, stabilizes the nodes that would otherwise oscillate given the repulsion and attraction parameters.

- The **Autostab Strength** parameter controls the strength of the auto-stabilization function; high values will cause the wobbly nodes to stabilize quickly.

- **Autostab sensibility** defines the extent to which the inertia parameter changes during the algorithm's execution.

- The **Gravity** parameter specifies the strength of an attraction of each node toward the center of the graph.

- ▸ The **Attraction Distrib.** parameter controls the distribution of attraction centers so that the graph looks balanced. There would be a central hub and spokes distributed toward the edges of the graph.

- ▸ **Adjust by Sizes** controls the overlapping of nodes; when checked, it will prevent the nodes from overlapping.

- ▸ The **Speed** parameter controls the speed of the algorithm; high values of this parameter (has to be greater than 0) will speed up the algorithm convergence at the cost of losing some of the precision.

The following image shows our graph with different repulsion strength parameters:

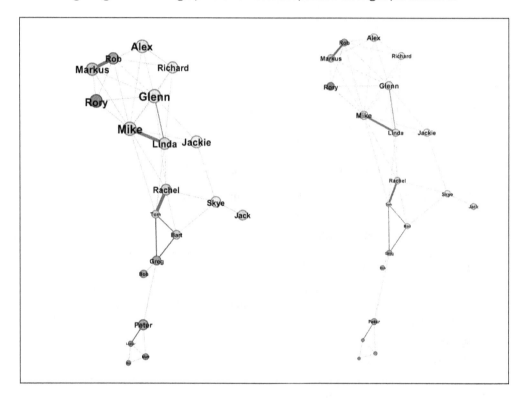

As you can see, both of the graphs have essentially the same shape, but the one on the right is much more dispersed—we cannot even read the names of the older generation.

The `Weight` parameter controls the size of the edge line. However, the small lines are now barely visible. You can control the thickness of the lines using the slider in the bottom-left corner of the graph window:

I also colored the edges differently depending on the type of the relationship. To do this, go to the graph controls panel and select **Partition**. From there, go to the **Edges** tab and select **relationship** from the drop-down:

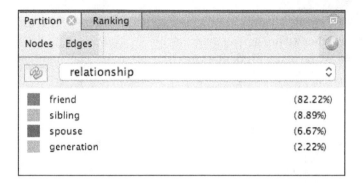

 Note that in the new version, **Ranking** and **Partition** have been merged into a tab called **Appearance**.

Once you apply the changes, you will see that the colors change on the graph. Note that now we have encoded the relationship not only with a color, but also with a weight. The final graph looks as follows:

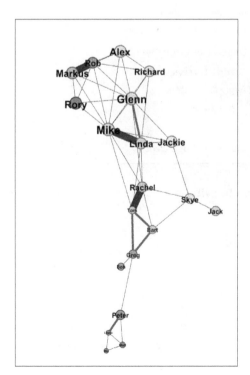

You can clearly see the hubs, delineating the age differences between the people in the social network.

There's more...

Let's now check whether we are getting the same results from Gephi as we were getting from NetworkX earlier. We go to the **Statistics** and **Filter** panel. Let's compare the graph density:

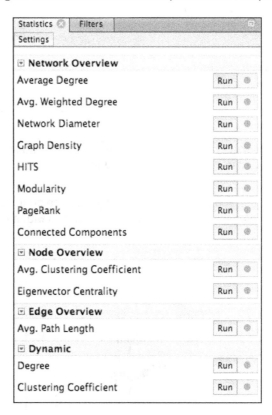

Click on **Run** next to **Graph Density**; in my case, it reads exactly the same value as before from NetworkX: 0.237.

Finally, let's explore some of the relationships in our data. We will use filters to achieve this. First, let's see who is married:

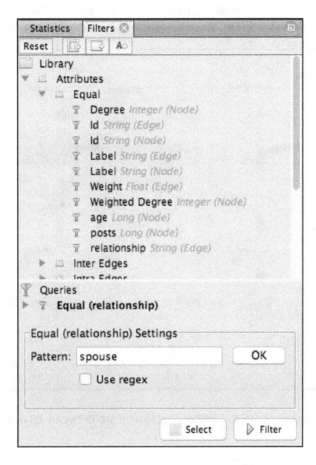

From **Library**, navigate to **Attributes | Equal** (we'll be selecting only the edges that are equal to `spouse`), select **relationship**, and drag it to **Queries**. The **Equal (relationship) Settings** window should appear below. In **Pattern**, type `spouse` and click on **OK**. Now, depending on whether you clicked **Select** or **Filter**, you will see a graph as follows (respectively):

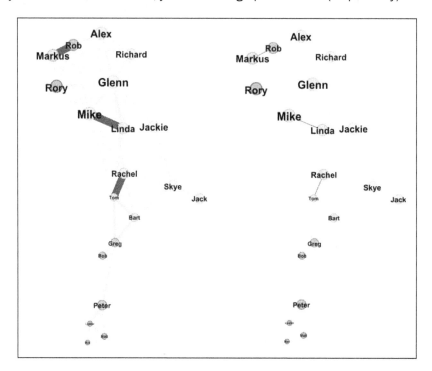

You can also stack up the filters. Let's filter people who are between 18 and 32 years of age and married:

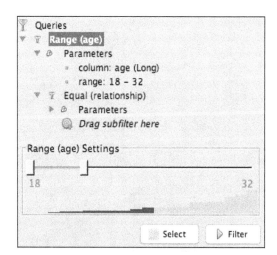

We start with using **Range** on the age filter. Using the slider, we select all the nodes with ages between 18 and 32. At the very bottom of the **Range** filter, you will see (not in the preceding image as it is already occupied by the **Equal** filter) a spot that will say **Drag subfilter here**; drag an **Equal** filter there for relationship and specify spouse. Now, if you click on the **Range (age)** filter and hit **Select**, you will see an image similar to the following one:

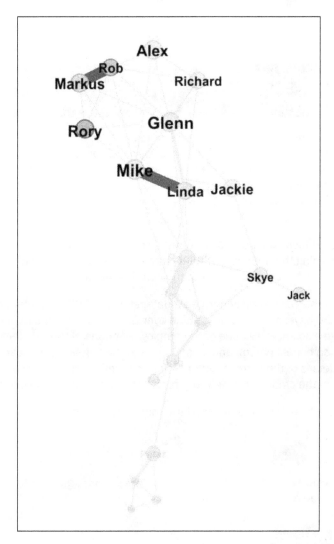

You can see that we are just selecting people younger than 32 and highlighting those who are married.

See also

- ▶ If you want to learn more, I strongly recommend you to explore the resources available at Gephi's website: `https://gephi.org/users/`
- ▶ Check out the following book: `https://www.packtpub.com/big-data-and-business-intelligence/network-graph-analysis-and-visualization-gephi`

Identifying people whose credit card details were stolen

Nowadays, being a victim of a fraud is not as remote a concept as people might think. While you can feel fairly secure online using double encryption and strong passwords, the old-school credit card information can be stolen relatively easily. With credit card frauds increasing at an alarming rate (`http://www.economist.com/news/finance-and-economics/21596547-why-america-has-such-high-rate-payment-card-fraud-skimming-top`), totaling to around $5.5 billion in 2012, it is not a laughing matter.

In this recipe, we will focus on a particular form of credit card fraud—buying from an online store. We are assuming that for some of those transactions (of a higher value), some retailers require the customers to call in and confirm their credit card details.

For the purpose of this recipe, we generated a dataset with 1,000 customers and 20 merchants. Over 50 days, our customers made over 225 K transactions for a total value of over $57 M. We also know that one of the employees of one of the retailers (merchant number 4) is not an honest person and, every now and then, steals credit card details from customers. The details of the cards are then being sold online and, when someone uses the card, the owners of the credit cards rightfully report the unauthorized payments.

In this recipe, you will learn how to extract data from a graph and find the victims of a fraud.

Getting ready

To execute this recipe, you will need `NetworkX`, `collections`, and `NumPy`. No other prerequisites are required.

How to do it...

We saved the data in a GraphML (gzipped) format in the `Data/Chapter08` folder. `NetworkX` makes it really easy to read the GraphML data even when it's packed in an archive (the `graph_fraudTransactions.py` file):

```
import networkx as nx
import numpy as np
```

```
import collections as c

# import the graph
graph_file = '../../Data/Chapter08/fraud.gz'
fraud = nx.read_graphml(graph_file)
```

How it works...

First, once we read the data, let's see what type of graph we are dealing with:

```
print('\nType of the graph: ', type(fraud))
```

As any given person could have made multiple transactions with any merchant, we are dealing with a directed graph with parallel edges, with each edge being a transaction. `NetworkX` confirms this:

```
Type of the graph:    <class 'networkx.classes.multidigraph.
MultiDiGraph'>
```

Let's confirm the number of nodes and edges:

```
# population and merchants
nodes = fraud.nodes()

nodes_population = [n for n in nodes if 'p_' in n]
nodes_merchants  = [n for n in nodes if 'm_' in n]

n_population = len(nodes_population)
n_merchants  = len(nodes_merchants)

print('\nTotal population: {0}, number of merchants: {1}' \
    .format(n_population, n_merchants))

# number of transactions
n_transactions = fraud.number_of_edges()
print('Total number of transactions: {0}' \
    .format(n_transactions))

# what do we know about a transaction
p_1_transactions = fraud.out_edges('p_1', data=True)
print('\nMetadata for a transaction: ',
    list(p_1_transactions[0][2].keys()))

# total value of all transactions
print('Total value of all transactions: {0}' \
    .format(np.sum([t[2]['amount']
        for t in fraud.edges(data=True)])))
```

First, we retrieve all the nodes from the graph. We know that all the nodes for people are prefixed by `p_` and by `m_` for merchants; we create lists of population and merchant nodes and check their lengths:

```
Total population: 1000, number of merchants: 20
```

The `.number_of_edges()` method returns the total number of edges in the graph; this is the number of all transactions:

```
Total number of transactions: 225037
```

Next, we check what metadata is available about the transactions: we retrieve a list of all transactions for `p_1` using the `.out_edges(...)` method. The method returns a list of all outgoing edges for `p_1` and all the metadata associated with them (the `data=True` parameter). As shown in the *Handling graph objects in Python with NetworkX* recipe, an element of this list is a tuple of three elements: (origin, destination, and metadata); the metadata element is a dictionary so we grab the keys of this element:

```
Metadata for a transaction:  ['type', 'disputed', 'time', 'amount']
```

In our graph, `type` is always a purchase, the `time` parameter provides a timestamp of when the transaction was made, and `amount` states the value of the transaction. The disputed flag tells us whether a transaction was disputed or not.

So, let's check how much people have spent online over the course of 50 days. We loop through all the edges and create a list with the values of all transactions. Finally, the `.sum(...)` method of `NumPy` sums all the elements of our list to provide the final value:

```
Total value of all transactions: 57273724
```

There's more...

Now that we know the basics about our graph, the first step in identifying the source of the scam is to identify the scammed customers:

```python
# identify customers with stolen credit cards
all_disputed_transactions = \
    [dt for dt in fraud.edges(data=True) if dt[2]['disputed']]

print('\nDISPUTED TRANSACTIONS')
print('Total number of disputed transactions: {0}' \
    .format(len(all_disputed_transactions)))
print('Total value of disputed transactions: {0}' \
    .format(np.sum([dt[2]['amount']
        for dt in all_disputed_transactions])))

# a list of people scammed
```

```
people_scammed = list(set(
    [p[0] for p in all_disputed_transactions]))

print('Total number of people scammed: {0}' \
    .format(len(people_scammed)))

# a list of all disputed transactions
print('All disputed transactions:')

for dt in sorted(all_disputed_transactions,
    key=lambda e: e[0]):
    print('({0}, {1}: {{time:{2}, amount:{3}}})'\
        .format(dt[0], dt[1],
        dt[2]['amount'], dt[2]['amount']))
```

First, we identify all the disputed transactions. We do so by checking the disputed flag while we loop through all the edges:

```
Total number of disputed transactions: 49
```

So, out of 225,037 transactions, 49 were fraudulent. It is not a huge number, but if we leave it unchecked and do not identify the source of the fraud, it can explode in the future. Also, the count does not account for undisputed transactions so this number might well be much higher.

Next, let's check how much was stolen. As with the total value of all the transactions, we loop through all the transactions that were disputed and sum their value:

```
Total value of disputed transactions: 14277
```

Altogether, over $14 K was stolen; this equates to around $290 a transaction.

We still do not know how many people were scammed. We loop through all the transactions retrieving all the people who had their money stolen. The set(...) method reduces the list to distinct values only (just like a mathematical set cannot have two same elements). We parse the set back to a list:

```
Total number of people scammed: 33
```

In total, 33 people had their credit cards stolen. This corresponds to 1.48 transactions per person and around $430 of value per person.

Let's look at those transactions (abbreviated).

As the .format(...) method uses {.} to indicate the places where to put the corresponding value in the template string, you need to use double braces {{}} to print literal {} (braces).

The top 10 disputed transactions are listed here:

```
All disputed transactions:
(p_114, m_12: {time:290, amount:290})
(p_123, m_12: {time:273, amount:273})
(p_154, m_2: {time:448, amount:448})
(p_164, m_3: {time:98, amount:98})
(p_224, m_2: {time:162, amount:162})
(p_272, m_2: {time:489, amount:489})
(p_276, m_3: {time:122, amount:122})
(p_325, m_2: {time:409, amount:409})
(p_389, m_12: {time:460, amount:460})
(p_389, m_2: {time:262, amount:262})
...
```

Finally, let's see how much each person lost:

```
# how much each person lost
transactions = c.defaultdict(list)

for p in all_disputed_transactions:
    transactions[p[0]].append(p[2]['amount'])

for p in sorted(transactions.items(),
    key=lambda e: np.sum(e[1]), reverse=True):
    print('Value lost by {0}: \t{1}'\
        .format(p[0], np.sum(p[1])))
```

We start by creating `.defaultdict(...)`. A `.defaultdict(...)` is a dictionary-like object. However, with a regular dictionary, if the value for each key is supposed to be a list and the key does not exist, you cannot `.append(...)` to the list as it does not yet exist. (If you try to, Python will complain and throw an exception.) With `.defaultdict(...)`, instead of throwing an exception, the structure first inserts a new key into the dictionary with an empty list and then appends the value to the newly created list. So, we can do `transactions[p[0]].append(p[2]['amount'])` instead of having to do the following:

```
for p in all_disputed_transactions:
    try:
        transactions[p[0]].append(p[2]['amount'])
    except:
        transactions[p[0]] = [p[2]['amount']]
```

Once we have all the transactions split into the affected people, we can print out the list of customers affected by the scammer (from the most affected to the least as we specify `reverse=True`):

```
Value lost by p_389:    1453
Value lost by p_721:    1383
Value lost by p_583:    878
Value lost by p_607:    750
Value lost by p_471:    675
Value lost by p_504:    581
Value lost by p_70:     519
Value lost by p_272:    489
Value lost by p_8:      486
Value lost by p_684:    484
Value lost by p_545:    477
Value lost by p_514:    463
Value lost by p_154:    448
Value lost by p_415:    410
Value lost by p_325:    409
Value lost by p_637:    365
Value lost by p_865:    361
Value lost by p_54:     356
Value lost by p_540:    343
Value lost by p_709:    342
Value lost by p_590:    328
Value lost by p_114:    290
Value lost by p_542:    282
Value lost by p_123:    273
Value lost by p_577:    224
Value lost by p_482:    215
Value lost by p_734:    197
Value lost by p_418:    163
Value lost by p_224:    162
Value lost by p_908:    134
Value lost by p_276:    122
Value lost by p_392:    117
Value lost by p_164:    98
```

So, we can see that the distribution is not that even; some of the people affected lost over $1,000 and seven people lost over $500. This is not a small dent to any family's budget and should be investigated.

Identifying those responsible for stealing the credit cards

Having identified the affected parties and seeing how much they lost, let's now find out who is responsible for this.

Getting ready

To execute this recipe, you will need `NetworkX`, `collections`, and `NumPy`. No other prerequisites are required.

How to do it...

In this recipe, we will attempt to find the merchant that all the affected parties shopped at before the first fraudulent transaction occurred (the `graph_fraudOrigin.py` file):

```python
import networkx as nx
import numpy as np
import collections as c

# import the graph
graph_file = '../../Data/Chapter08/fraud.gz'
fraud = nx.read_graphml(graph_file)

# identify customers with stolen credit cards
people_scammed = c.defaultdict(list)

for (person, merchant, data) in fraud.edges(data=True):
    if data['disputed']:
        people_scammed[person].append(data['time'])

print('\nTotal number of people scammed: {0}' \
    .format(len(people_scammed)))

# what was the time of the first disputed transaction for each
# scammed person
stolen_time = {}

for person in people_scammed:
    stolen_time[person] = \
        np.min(people_scammed[person])

# let's find the common merchants for all those scammed
merchants = c.defaultdict(list)
for person in people_scammed:
    edges = fraud.out_edges(person, data=True)

    for (person, merchant, data) in edges:
        if  stolen_time[person] - data['time'] <= 1 and \
```

```
            stolen_time[person] - data['time'] >= 0:

        merchants[merchant].append(person)

merchants = [(merch, len(set(merchants[merch])))
    for merch in merchants]

# print the top 5 merchants
print('\nTop 5 merchants where people made purchases')
print('shortly before their credit cards were stolen')
print(sorted(merchants, key=lambda e: e[1], reverse=True)[:5])
```

How it works...

First, we read in the data in a similar fashion as before. Next, we create a list of scammed_
people but in a slightly different way than in the most recent recipe. Instead of creating a list,
we want to find the time of the first disputed transaction. So we use .defaultdict(...)
again, loop through all the disputed transactions, and create a dictionary where, for each
person, we will get a list of times of the disputed transactions reported by the person.

We do this as we want to check all the transactions prior to the first disputed transaction; it
is physically impossible for the fraudster to commit the crime without first stealing the credit
card details.

With the list of all the affected customers, we check when the first disputed transaction
occurred by looping through all the elements of the scammed_people dictionary and
finding the minimum time of all the disputed transactions. We store this in the stolen_time
dictionary.

Now it is time to check the transactions prior to the first disputed one. In the for loop, we first
select all the transactions for each affected person and then loop through all the transactions
that occurred before the first disputed one.

In this recipe, we look only one day back as we check stolen_time[person] -
data['time'] <= 1 and stolen_time[person] - data['time'] >= 0.

However, this time frame might be shrunk or extended, depending on when you first hit the
common merchant for all the affected customers. In our case, we hit it first at one: all of the
33 affected customers shopped with the same merchant no later than one day prior to the
first fraudulent transaction that occurred.

With the list of transactions, let's find the merchants. We are again using the set(...)
operation to select only the distinct merchants for each customer (as any of them could have
had multiple transactions with the same merchant over this period of time) and count how
many elements our set has.

Finally, let's check who is the winner:

```
Top 5 merchants where people made purchases
shortly before their credit cards were stolen
[('m_4', 33), ('m_2', 16), ('m_3', 14), ('m_12', 9), ('m_6', 8)]
```

As you can see, all of the affected customers shopped at merchant m_4 shortly before their credit cards were stolen. As we generated the data, we know this to be true. However, in a real world, these top five merchants would be contacted and an investigation for a black sheep among their employees would be conducted.

See also

To develop these recipes, we adapted the method from graph databases, namely, Neo4j. You can read more about Neo4j and the fraud detection with graph databases at https:// linkurio.us/stolen-credit-cards-and-fraud-detection-with-neo4j/.

9

Natural Language Processing

In this chapter, you will learn the following recipes:

- ▶ Reading raw text from the Web
- ▶ Tokenizing and normalizing text
- ▶ Identifying parts of speech, handling n-grams, and recognizing named entities
- ▶ Identifying the topic of an article
- ▶ Identifying the sentence structure
- ▶ Classifying movies based on their reviews

Introduction

Modeling based on structured data gathered via a controlled experiment (as we were doing in previous chapters) is relatively straightforward. However, in the real world, we rarely deal with structured data. This is especially true when it comes to understanding human-generated feedback or analyzing an article in a newspaper.

Natural Language Processing (**NLP**) is a discipline of computer science, statistics, and linguistics that aims at processing human language (I consciously did not use the word, understanding) and extracting features that can be used in modeling. Using NLP concepts, among other tasks, we can find the most occurring words in a text in order to roughly identify the topic of such a body of text, identify names of people and places, find objects and subjects in a sentence, or analyze the sentiment of someone's feedback.

In this set of recipes, we will be using two datasets. We will read the first one off the Seattle Times website—the Obama moves to require *background* checks for more gun sales article by Josh Lederman (`http://www.seattletimes.com/nation-world/obama-starts-2016-with-a-fight-over-gun-control/`, accessed 1/04/2016).

The other dataset consists of 50,000 movie reviews that were preprocessed (and de-identified) by A. L. Mass et al.; the full dataset can be found at `http://ai.stanford.edu/~amaas/data/sentiment/`. It was published in *Andrew L. Maas, Raymond E. Daly, Peter T. Pham, Dan Huang, Andrew Y. Ng*, and *Christopher Potts (2011), Learning Word Vectors for Sentiment Analysis, The 49th Annual Meeting of the Association for Computational Linguistics (ACL 2011)*.

Out of all the 50,000 movie reviews, we only selected 2,000 positive and 2,000 negative from the training batch and 2,000 positive and 2,000 negative from the testing batch.

Reading raw text from the Web

Most of the times, the free-form text can be found in text files; in this recipe, we will not be teaching you how to do that as we have already presented many ways of doing so. (Refer to the set of recipes in *Chapter 1, Preparing the Data*.)

One way of reading a file that we have not explored yet will be discussed in the next recipe.

Many times, however, we need to read data straight from the web: we might want to analyze a blog post, scrape an article, or analyze Facebook or Twitter posts. While Facebook and Twitter offer **Application Programming Interfaces** (**APIs**) that normally return answers in XML or JSON formats, processing HTML files is not as straightforward.

In this recipe, you will learn how to access a web page, read its content, and process it.

Getting ready

To execute this recipe, you will need `urllib`, `html5lib`, and `Beautiful Soup`.

Urllib comes with Python 3 (`https://docs.python.org/3/library/urllib.html`). If, however, your configuration does not have `Beautiful Soup`, it is easy to install.

In your command line (assuming that you are using Anaconda and Python 3.5), issue the following statement:

conda install beautifulsoup4

Additionally, in order to parse an HTML file with Beautiful Soup, we need to install html5lib; with Anaconda, it is again extremely easy:

conda install html5lib

No other prerequisites are required.

How to do it...

The process of accessing websites using urllib has changed slightly between Python 2.x and Python 3.x: urllib2 (available in Python 2.x) has been split into urllib.request, urllib.error, urllib.parse, and urllib.robotparser.

 Check https://docs.python.org/2/library/urllib2.html for more information.

In this recipe, we will be using urllib.request (the nlp_read.py file):

```
import urllib.request as u
import bs4 as bs

# link to the article at The Seattle Times
st_url = 'http://www.seattletimes.com/nation-world/obama-starts-2016-
with-a-fight-over-gun-control/'

# read the contents of the webpage
with u.urlopen(st_url) as response:
    html = response.read()

# using beautiful soup -- let's parse the content of the HTML
read = bs.BeautifulSoup(html, 'html5lib')

# find the article tag
article = read.find('article')

# find all the paragraphs of the article
all_ps = article.find_all('p')
```

```
# object to hold the body of text
body_of_text = []

# get the tile
body_of_text.append(read.title.get_text())

# put all the paragraphs to the body of text list
for p in all_ps:
    body_of_text.append(p.get_text())

# we don't need some of the parts at the bottom of the page
body_of_text = body_of_text[:24]

# let's see what we got
print('\n'.join(body_of_text))

# and save it to a file
with open('../../Data/Chapter09/ST_gunLaws.txt', 'w') as f:
    f.write('\n'.join(body_of_text))
```

How it works...

As always, we start with importing the necessary modules; in this case, `urllib` and `Beautiful Soup`.

The link to the Seattle Times article that we will use is stored in the `st_url` object. The `.urlopen(...)` method of `urllib` opens the specified URL.

We use the already familiar `with(...)` as `...` construct as it handles closing the connection properly when we do not use it anymore. You could, of course, do it as follows:

```
local_filename, headers = \
    urllib.request.urlretrieve(st_url)
html = open(local_filename)
html.close()
```

The `.read()` method on the response object reads the whole content of the web page. If you were to print it out, you would see something as follows (abbreviated for obvious reasons):

```
b'<!DOCTYPE html>\n<!--[if lt IE 10]>          <html class="no-
js lt-ie10 no-support" lang="en-US" itemscope itemtype="http://
schema.org/Article" > <![endif]-->\n<!--[if gt IE 9]><!-->    <html
class="no-js"          lang="en-US" itemscope itemtype="http://
schema.org/Article" > <!--<![endif]-->\n<head>\n      \n  <meta
http-equiv="X-UA-Compatible" content="IE=edge,chrome=1"/>\n
<meta charset="UTF-8"><script type="text/javascript">(window.
NREUM||(NREUM={})).loader_config={xpid:"XAcAVFdRGwIFUVhQBAIB"};wi
ndow.NREUM||(NREUM={}),__nr_require=function(t,e,n){function r(n)
{if(!e[n]){var o=e[n]={exports:{}};t[n][0].call(o.exports,function(e)
{var o=t[n][1][e];return r(o||e)},o,o.exports)}return e[n].exports}
if("function"==typeof __nr_require)return __nr_require;for(var
o=0;o<n.length;o++)r(n[o]);return r}({QJf3ax:[function(t,e){function
n(t){function e(e,n,a){t&&t(e,n,a),a||(a={});for(var c=s(e),f=c.
length,u=i(a,o,r),d=0;f>d;d++)c[d].apply(u,n);return u}function a(t,e)
{f[t]=s(t).concat(e)}function s(t){return f[t]||[]}function c(){return
n(e)}var f={};return{on:a,emit:e,create:c,listeners:s,_events:f}}
function r(){return{}}var o="nr@context",i=t("gos");e.exports=n()},{
gos:"7eSDFh"}],ee:[function(t,e){e.exports=t("QJf3ax")},{}],3:[funct
ion(t){function e(t){try{i.console&&console.log(t)}catch(e){}}var n,
r=t("ee"),o=t(1),i={};try{n=localStorage.getItem("__nr_flags").split
(","),console&&"function"==typeof console.log&&(i.console=!0,-1!==n.
indexOf("dev")&& ...
```

This is the web page represented literally as a text. This is not really what we need for our text analysis.

`Beautiful Soup` to the rescue! The `BeautifulSoup(...)` method accepts HTML or XML text as its first argument. The second argument specifies what parser to use.

For a list of available parsers, check `http://www.crummy.com/software/BeautifulSoup/bs4/doc/#specifying-the-parser-to-use`.

After parsing, it is (somewhat) more readable:

```
<!DOCTYPE html>
<!--[if lt IE 10]>          <html class="no-js lt-ie10 no-support"
lang="en-US" itemscope itemtype="http://schema.org/Article" >
<![endif]--><!--[if gt IE 9]><!--><html class="no-js" itemscope=""
itemtype="http://schema.org/Article" lang="en-US"><!--<![endif]--
><head>

    <meta content="IE=edge,chrome=1" http-equiv="X-UA-Compatible"/>
    <meta charset="utf-8"/>
```

```
    <meta content="width=device-width, initial-scale=1.0, user-
scalable=0" name="viewport"/>
    <title>
       Obama moves to require background checks for more gun sales | The
Seattle Times     </title>
    <meta content="Although Obama can't unilaterally change gun laws,
the president is hoping that beefing up enforcement of existing laws
can prevent at least some gun deaths in a country rife with them."
name="description"/>
    <!-- Google+ -->
    <meta content="Obama moves to require background checks for more gun
sales" itemprop="name"/>
    <meta content="Obama moves to require background checks for more gun
sales" itemprop="headline"/>
    <meta content="2016-01-04 00:50:06" itemprop="datePublished"/>
   ...
```

However, we do not use `Beautiful Soup` to print the results to the screen. The `BeautifulSoup` object is internally represented as a hierarchical bag of all tags found in the document.

 You could also think of the `BeautifulSoup` object as a tree.

Given the preceding output, the good thing is that you can search for all the tags that are present in the HTML or XML file. The newer web pages (compliant with HTML5) have additional tags that help publish content on the web in an easier way.

 The list of all new elements can be found here:
`http://www.w3.org/TR/html5-diff/#new-elements`

In our case, we first find and extract the article tag; this narrows our search to what we really want out of the Seattle Times page:

```
<article class="post-9873571 post type-post status-publish format-
standard hentry h-entry" id="post-9873571">
<header class="article-header ">
<ul class="article-slug">
       <li><a href="/seattle-news/crime/" rel="tag">Crime</a></
li><li><a href="/nation-world/nation/" rel="tag">Nation</a></li><li><a
href="/nation-world/" rel="tag">Nation &amp; World</a></li><li><a
href="/nation-world/nation-politics/" rel="tag">Nation &amp; World
Politics</a></li>        </ul>
             <h1 class="article-title p-name entry-title">
       Obama moves to require background checks for more gun sales
   </h1>
```

```
        <div class="article-share title" data-utm="article_title">
        <a class="social-share"><i class="icon-facebook large"></i></a>
        <a class="social-share"><i class="icon-mail large"></i></a>
        <a class="social-share"><i class="icon-twitter large"></i></a>
        <div id="mobile-sponsor-ad"></div>
    </div>
    <div class="article-dateline">
            <time class="line published dt-published"
datetime="2016-01-04 00:50:06">Originally published January 4, 2016 at
12:50 am</time>
            <time class="line update updated dt-updated"
datetime="2016-01-04 21:40:10">
            Updated January 4, 2016 at 9:40 pm          </time>
            </div>
    . . .
```

So, now we are only looking at the content of the article and discard most of the glue that creates the web page. If you open the web page itself, you can start to see familiar sentences, as follows:

Obama moves to require background checks for more gun sales.

Originally published January 4, 2016 at 12:50 am.

We are on the right track. Scrolling a little lower, we get to more familiar sentences from the body of the article:

```
<p>WASHINGTON (AP) — President Barack Obama moved Monday to expand
background checks to cover more firearms sold at gun shows, online
and anywhere else, aiming to curb a scourge of gun violence despite
unyielding opposition to new laws in Congress.</p>
<p>Obama's plan to broaden background checks forms the centerpiece
of a broader package of gun control measures the president plans to
take on his own in his final year in office. Although Obama can't
unilaterally change gun laws, the president is hoping that beefing up
enforcement of existing laws can prevent at least some gun deaths in a
country rife with them.</p>
    . . .
```

Now it is becoming apparent that all the paragraphs of the article are enclosed in the `<p>` paragraph tags. So, we use the `.find_all('p')` method to extract them all.

Next, we add the title to the `body_of_text` list. We use the `.get_text()` method that extracts just the text from the tag; otherwise, we would get the text with the tags:

```
<title>
    Obama moves to require background checks for more gun sales | The
Seattle Times    </title>
```

We use the same method to extract the text from all the paragraphs. What you should see when we print it out to the screen is presented (in an abbreviated form) here:

```
Obama moves to require background checks for more gun sales | The
Seattle Times
Although Obama can't unilaterally change gun laws, the president is
hoping that beefing up enforcement of existing laws can prevent at
least some gun deaths in a country rife with them.
WASHINGTON (AP) — President Barack Obama moved Monday to expand
background checks to cover more firearms sold at gun shows, online
and anywhere else, aiming to curb a scourge of gun violence despite
unyielding opposition to new laws in Congress.
Obama's plan to broaden background checks forms the centerpiece of a
broader package of gun control measures the president plans to take on
his own in his final year in office. Although Obama can't unilaterally
change gun laws, the president is hoping that beefing up enforcement
of existing laws can prevent at least some gun deaths in a country
rife with them.
...
```

Now, that's more like something we can work with. Finally, we save the text in a file.

Tokenizing and normalizing text

Extracting the contents of the page is just the first step. Before we get to the fun part of analyzing what the article contains (or, if you looked at blog posts, what they are about), we need to split the whole article into sentences and further into words.

Having done so, we would still face another issue; in any of the text, we would see sentences in different tenses, people using the passive voice, or some rarely seen grammatical constructs. For the purpose of extracting the topic or analyzing the sentiment, we do not really need to see words `said` and `says` separately—the word `say` would be enough. Thus, we will also be looking at normalizing the text, that is, bringing all the different versions of the same word to some common form.

Getting ready

To execute this recipe, all you need is the **Natural Language Toolkit** (**NLTK**). Before we start, however, you need to make sure that the NLTK module is present on your machine. If you are using Anaconda, this is simple; you can use the following (on Mac or Linux):

```
conda list | grep nltk
```

You can also use the following (on Windows):

```
conda install nltk
```

If you see something similar to the following from executing the first command, you are ready to go:

```
nltk                     3.1                     py34_0     defaults
```

If you get the following from executing the second one, you are ready to go:

```
Fetching package metadata: ....
Solving package specifications: ....................
# All requested packages already installed.
# packages in environment at /Users/drabast/anaconda:
#
nltk                     3.1                     py34_0
```

This, however, only makes sure that the code for NLTK is present on your machine. In order to make full use of the package, we need to download the data used by certain parts of the module. This can be done by executing the following Python script (the `nlp_download.py` file):

```
import nltk
nltk.download()
```

Once it starts executing, you will see the following window pop up on your screen:

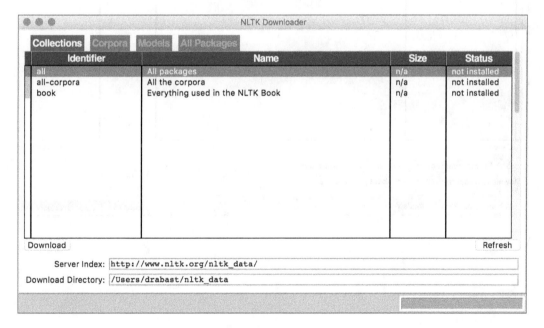

By default, it highlights all and I recommend that you keep it that way. The other options would download only the corpora (all-corpora) or everything needed to follow the NLTK book.

 The NLTK book can be accessed at `http://www.nltk.org/book/` and I highly recommend reading it if you are interested in the NLP topic.

Following this, click on the **Download** button. The download process will start and you should be observing a window similar to the following one for quite some time; it took me around 45-50 minutes to download all the corpora and other things. Be patient.

 If your download fails and you cannot download it all, you can just install the `punkt` model. Navigate to the **All Packages** tab and select **punkt**.

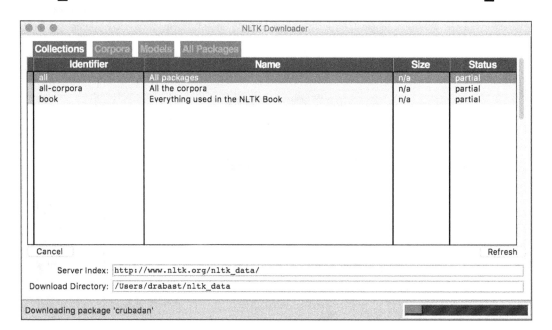

Once the process finishes, you shall see a screen like this with everything highlighted in green:

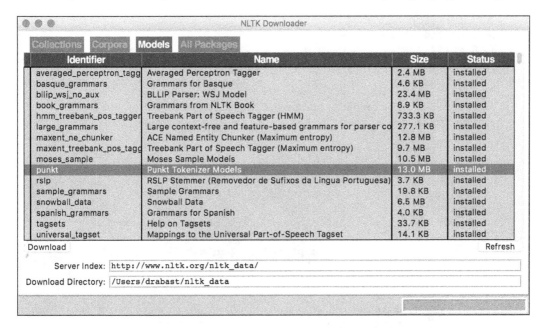

No other prerequisites are required.

How to do it...

Now that we have everything that we need to move on, let's tokenize our text and normalize it (the `nlp_tokenize.py` file):

```
import nltk

# read the text
guns_laws = '../../Data/Chapter09/ST_gunLaws.txt'

with open(guns_laws, 'r') as f:
    article = f.read()

# load NLTK modules
sentencer = nltk.data.load('tokenizers/punkt/english.pickle')
tokenizer = nltk.RegexpTokenizer(r'\w+')
stemmer = nltk.PorterStemmer()
lemmatizer = nltk.WordNetLemmatizer()

# split the text into sentences
sentences = sentencer.tokenize(article)
words = []
stemmed_words = []
```

```
lemmatized_words = []

# and for each sentence
for sentence in sentences:
    # split the sentence into words
    words.append(tokenizer.tokenize(sentence))

    # stem the words
    stemmed_words.append([stemmer.stem(word)
        for word in words[-1]])

    # and lemmatize them
    lemmatized_words.append([lemmatizer.lemmatize(word)
        for word in words[-1]])

# and save the results to files
file_words  = '../../Data/Chapter09/ST_gunLaws_words.txt'
file_stems  = '../../Data/Chapter09/ST_gunLaws_stems.txt'
file_lemmas = '../../Data/Chapter09/ST_gunLaws_lemmas.txt'

with open(file_words, 'w') as f:
    for w in words:
        for word in w:
            f.write(word + '\n')

with open(file_stems, 'w') as f:
    for w in stemmed_words:
        for word in w:
            f.write(word + '\n')

with open(file_lemmas, 'w') as f:
    for w in lemmatized_words:
        for word in w:
            f.write(word + '\n')
```

How it works...

First, we read in the text from the ST_gunLaws.txt file. As our text file has almost no structure (apart from the paragraphs being separated by \n), we can really read it raw from the file; the .read() method does just this for us.

Next, we load all the necessary NLTK modules. The sentencer is a punkt sentence tokenizer. The tokenizer uses an unsupervised algorithm to find the places where sentences start and finish. A naïve approach to sentence tokenizing would be to look for full stop characters, '.'. This, however, would not handle situations where we have some abbreviated words in the middle of the sentence that require the full stop character, for example, Dr. or someone's name, like Michael D. Brown.

More on the punkt sentence tokenizer can be found in NTLK's documentation: `http://www.nltk.org/api/nltk.tokenize.html#module-nltk.tokenize.punkt`

In this recipe, we use the regular expressions word tokenizer (so, for now, we do not need to handle the punctuation). The `.RegexpTokenizer(...)` takes the regular expression as its argument; in our case, we are only interested in distinct words hence we use `'\w+'`. The drawback of using this tokenizer is that it will handle the words with apostrophes wrongly, that is, the word `don't` will be tokenized into `['don', 't']`.

For a full list of regular expressions available in Python, check `https://docs.python.org/3/library/re.html`.

The stemmer object removes the ending of words according to a specific algorithm (or set of rules). We use `.PorterStemmer()` for the purpose of word normalization; the stemmer strips `ing` from `reading` and outputs `read`. However, the stemmer can be really crude for some words, for example, the word `president` would be stemmed to `presid` as the morphological base for the word `president` would be `preside` (which, if stemmed, would also be presented as `presid`).

The Porter Stemmer has been ported from ANSI C. The stemmer was originally proposed by C.J. van Rijsbergen, S.E. Robertson, and M.F. Porter, developed as a part of a larger IR project. The An algorithm for suffix stripping paper was later published by M.F. Porter in 1980 in the Program journal. Check `http://tartarus.org/~martin/PorterStemmer/` if you are interested.

The purpose of lemmatization is really related to that of stemming as it also aims at normalizing the text; for instance, words `are` and `is` would be normalized to `be`. The difference between a stemmer and lemmatizer is that the former (as explained previously) uses an algorithm to strip the ending of words (sometimes in very drastic ways) while the lemmatizer would use a large vocabulary to come up with the morphological base for the word.

In this recipe, we use the WordNet lemmatizer. More on this lemmatizer can be found at `http://www.nltk.org/api/nltk.stem.html#module-nltk.stem.wordnet`.

Having created all the necessary objects, we move to extracting sentences from the text. After passing through the punkt sentence tokenizer, we get a list of 35 sentences:

```
["Obama moves to require background checks for more gun sales | The
Seattle Times    \nAlthough Obama can't unilaterally change gun laws,
the president is hoping that beefing up enforcement of existing laws
can prevent at least some gun deaths in a country rife with them.",
'WASHINGTON (AP) — President Barack Obama moved Monday to expand
background checks to cover more firearms sold at gun shows, online
and anywhere else, aiming to curb a scourge of gun violence despite
unyielding opposition to new laws in Congress.',
'Obama's plan to broaden background checks forms the centerpiece of a
broader package of gun control measures the president plans to take on
his own in his final year in office.',
'Although Obama can't unilaterally change gun laws, the president is
hoping that beefing up enforcement of existing laws can prevent at
least some gun deaths in a country rife with them.', ...]
```

Next, we loop through all the sentences and tokenize all the words in the sentence using the tokenizer and append them to the words list. The words list is effectively a list of lists:

```
[['Obama', 'moves', 'to', 'require', 'background', 'checks', 'for',
'more', 'gun', 'sales', 'The', 'Seattle', 'Times', 'Although',
'Obama', 'can', 't', 'unilaterally', 'change', 'gun', 'laws', 'the',
'president', 'is', 'hoping', 'that', 'beefing', 'up', 'enforcement',
'of', 'existing', 'laws', 'can', 'prevent', 'at', 'least', 'some',
'gun', 'deaths', 'in', 'a', 'country', 'rife', 'with', 'them'],
['WASHINGTON', 'AP', 'President', 'Barack', 'Obama', 'moved',
'Monday', 'to', 'expand', 'background', 'checks', 'to', 'cover',
'more', 'firearms', 'sold', 'at', 'gun', 'shows', 'online', 'and',
'anywhere', 'else', 'aiming', 'to', 'curb', 'a', 'scourge', 'of',
'gun', 'violence', 'despite', 'unyielding', 'opposition', 'to', 'new',
'laws', 'in', 'Congress'], ['Obama', 's', 'plan', 'to', 'broaden',
'background', 'checks', 'forms', 'the', 'centerpiece', 'of', 'a',
'broader', 'package', 'of', 'gun', 'control', 'measures', 'the',
'president', 'plans', 'to', 'take', 'on', 'his', 'own', 'in', 'his',
'final', 'year', 'in', 'office'], ['Although', 'Obama', 'can', 't',
'unilaterally', 'change', 'gun', 'laws', 'the', 'president', 'is',
'hoping', 'that', 'beefing', 'up', 'enforcement', 'of', 'existing',
'laws', 'can', 'prevent', 'at', 'least', 'some', 'gun', 'deaths',
'in', 'a', 'country', 'rife', 'with', 'them'], ...]
```

Note that the word can't has been parsed into can and t—as expected from .RegexpTokenizer(...).

Now that each sentence has been split into words, let's do stemming.

 Just as a reminder: the words[-1] syntax means that we're selecting only the last element of that list (that we just appended in the preceding command).

After stemming, we get the following list:

```
[['Obama', 'move', 'to', 'requir', 'background', 'check', 'for',
'more', 'gun', 'sale', 'The', 'Seattl', 'Time', 'Although', 'Obama',
'can', 't', 'unilater', 'chang', 'gun', 'law', 'the', 'presid',
'is', 'hope', 'that', 'beef', 'up', 'enforc', 'of', 'exist', 'law',
'can', 'prevent', 'at', 'least', 'some', 'gun', 'death', 'in', 'a',
'countri', 'rife', 'with', 'them'], ['WASHINGTON', 'AP', 'Presid',
'Barack', 'Obama', 'move', 'Monday', 'to', 'expand', 'background',
'check', 'to', 'cover', 'more', 'firearm', 'sold', 'at', 'gun',
'show', 'onlin', 'and', 'anywher', 'els', 'aim', 'to', 'curb', 'a',
'scourg', 'of', 'gun', 'violenc', 'despit', 'unyield', 'opposit',
'to', 'new', 'law', 'in', 'Congress'], ['Obama', 's', 'plan', 'to',
'broaden', 'background', 'check', 'form', 'the', 'centerpiec',
'of', 'a', 'broader', 'packag', 'of', 'gun', 'control', 'measur',
'the', 'presid', 'plan', 'to', 'take', 'on', 'hi', 'own', 'in',
'hi', 'final', 'year', 'in', 'offic'], ['Although', 'Obama', 'can',
't', 'unilater', 'chang', 'gun', 'law', 'the', 'presid', 'is',
'hope', 'that', 'beef', 'up', 'enforc', 'of', 'exist', 'law',
'can', 'prevent', 'at', 'least', 'some', 'gun', 'death', 'in', 'a',
'countri', 'rife', 'with', 'them']]
```

You can now see the stemming in its full crudeness; words such as `require` were stemmed into `requir` or `Seattle` was stemmed into `Seattl`. However, words such as `moves` or `checks` were stemmed into `move` or `check` as expected. Let's see what we get from lemmatization:

```
[['Obama', 'move', 'to', 'require', 'background', 'check', 'for',
'more', 'gun', 'sale', 'The', 'Seattle', 'Times', 'Although',
'Obama', 'can', 't', 'unilaterally', 'change', 'gun', 'law', 'the',
'president', 'is', 'hoping', 'that', 'beefing', 'up', 'enforcement',
'of', 'existing', 'law', 'can', 'prevent', 'at', 'least', 'some',
'gun', 'death', 'in', 'a', 'country', 'rife', 'with', 'them'],
['WASHINGTON', 'AP', 'President', 'Barack', 'Obama', 'moved',
'Monday', 'to', 'expand', 'background', 'check', 'to', 'cover',
'more', 'firearm', 'sold', 'at', 'gun', 'show', 'online', 'and',
'anywhere', 'else', 'aiming', 'to', 'curb', 'a', 'scourge', 'of',
'gun', 'violence', 'despite', 'unyielding', 'opposition', 'to', 'new',
'law', 'in', 'Congress'], ['Obama', 's', 'plan', 'to', 'broaden',
'background', 'check', 'form', 'the', 'centerpiece', 'of', 'a',
'broader', 'package', 'of', 'gun', 'control', 'measure', 'the',
'president', 'plan', 'to', 'take', 'on', 'his', 'own', 'in', 'his',
'final', 'year', 'in', 'office'], ['Although', 'Obama', 'can', 't',
'unilaterally', 'change', 'gun', 'law', 'the', 'president', 'is',
'hoping', 'that', 'beefing', 'up', 'enforcement', 'of', 'existing',
'law', 'can', 'prevent', 'at', 'least', 'some', 'gun', 'death', 'in',
'a', 'country', 'rife', 'with', 'them']]
```

It looks like the lemmatizer removed -d from `required` properly and also handled the word `Seattle` properly. However, words in the past tense form were not handled properly; we can still see words such as `moved` or `sold`. In an ideal world, these should have been parsed to `move` and `sell` to truly normalize the text. The word `sold`, however, has not been parsed properly by either the stemmer or lemmatizer.

It is really up to you as to what you require for your analysis. Just bear in mind that the lemmatizer will most likely be slower than the stemmer as it has to look up the word in the dictionary.

As a last step, we save all the words in a file so that you can easily compare the two methods, as given the limited space we have here, we would not be able to do so.

See also

You can also see how others are doing the sentence and word tokenization:

`http://textminingonline.com/dive-into-nltk-part-ii-sentence-tokenize-and-word-tokenize`

Identifying parts of speech, handling n-grams, and recognizing named entities

One of the first things that you might want to look at is recognizing parts of speech for a word; it is really fundamental to understand in a sentence that the word checks is a verb or noun.

This, as useful as it is, will not help you handle bigrams (or, more generally, n-grams): clusters of words that, if analyzed separately (in a certain context), would lead to improper understanding of the text. For example, consider a phrase *neural networks* in an article on machine learning and, more specifically, an application of neural networks to control packet scheduling and routing in a local network. In the same article, these two words (`neural` and `networks`) can occur on their own with, to some degree, different meanings.

Finally, reading an article on politics at a recent meeting of the heads of states, we might encounter the word `President` quite frequently; what would be more interesting to understand is how many times do the words `President Obama` and `President Putin` occur in the text. We will show you how you might approach this issue in the following recipe.

Getting ready

To execute this recipe, you will need NLTK and the regular expressions module from Python. Both of these should already be available on your machine. No other prerequisites are required.

How to do it...

Tagging parts of speech is really simple with NLTK. Identifying named entities requires somewhat more work (the `nlp_pos.py` file):

```python
import nltk
import re

def preprocess_data(text):
    global sentences, tokenized
    tokenizer = nltk.RegexpTokenizer(r'\w+')

    sentences =  nltk.sent_tokenize(text)
    tokenized = [tokenizer.tokenize(s) for s in sentences]

# import the data
guns_laws = '../../Data/Chapter09/ST_gunLaws.txt'

with open(guns_laws, 'r') as f:
    article = f.read()

# chunk into sentences and tokenize
sentences = []
tokenized = []
words = []

preprocess_data(article)

# part-of-speech tagging
tagged_sentences = [nltk.pos_tag(s) for s in tokenized]

print(tagged_sentences)

# extract named entities -- naive approach
named_entities = []

for sentence in tagged_sentences:
    for word in sentence:
        if word[1] == 'NNP' or word[1] == 'NNPS':
            named_entities.append(word)

named_entities = list(set(named_entities))

print('Named entities -- simplistic approach:')
```

```
print(named_entities)

# extract names entities -- regular expressions approach
named_entities = []
tagged = []

pattern = 'ENT: {<DT>?(<NNP|NNPS>)+}'

# use regular expressions parser
tokenizer = nltk.RegexpParser(pattern)

for sent in tagged_sentences:
    tagged.append(tokenizer.parse(sent))

for sentence in tagged:
    for pos in sentence:
        if type(pos) == nltk.tree.Tree:
            named_entities.append(pos)

named_entities = list(set([tuple(e) for e in named_entities]))

print('\nNamed entities using regular expressions:')
print(named_entities)
```

How it works...

As usual, we start with importing the necessary modules and reading in the text that we want to analyze.

We prepared the `preprocess_data(...)` method so that we can automate sentence and word tokenization. Instead of using the Punkt sentence tokenizer, we let NLTK handle it for us: we use the `.sent_tokenize(...)` method that (under the hood) calls the Punkt sentence tokenizer anyway. For word tokenization, we still use the regular expressions tokenizer.

With our text split into sentences and sentences into words, we can now assign the parts of speech tags to each word. The `.pos_tag(...)` method looks at the whole sentence represented as an ordered list of tokenized words and returns a list of tuples; each tuple consists of the word itself and the assigned part of speech.

The idea behind assigning parts of speech at the sentence level is to infer (from the context) the correct part of speech for the word. For instance, the part of speech for the word `moves` would be different in the following two sentences:

"He `moves` effortlessly."

"His dancing `moves` are effortless."

In the first sentence, the word would be a verb while in the second sentence the word would be classified as a plural noun.

Let's see what we get:

```
[[('Obama', 'NNP'), ('moves', 'VBZ'), ('to', 'TO'), ('require', 'VB'),
('background', 'NN'), ('checks', 'NNS'), ('for', 'IN'), ('more',
'JJR'), ('gun', 'JJ'), ('sales', 'NNS'), ('The', 'DT'), ('Seattle',
'NNP'), ('Times', 'NNP'), ('Although', 'IN'), ('Obama', 'NNP'),
('can', 'MD'), ('t', 'VB'), ('unilaterally', 'RB'), ('change', 'JJ'),
('gun', 'NN'), ('laws', 'NNS'), ('the', 'DT'), ('president', 'NN'),
('is', 'VBZ'), ('hoping', 'VBG'), ('that', 'IN'), ('beefing', 'VBG'),
('up', 'RP'), ('enforcement', 'NN'), ('of', 'IN'), ('existing',
'VBG'), ('laws', 'NNS'), ('can', 'MD'), ('prevent', 'VB'), ('at',
'IN'), ('least', 'JJS'), ('some', 'DT'), ('gun', 'JJ'), ('deaths',
'NNS'), ('in', 'IN'), ('a', 'DT'), ('country', 'NN'), ('rife', 'NN'),
('with', 'IN'), ('them', 'PRP')], [('WASHINGTON', 'NNP'), ('AP',
'NNP'), ('President', 'NNP'), ('Barack', 'NNP'), ('Obama', 'NNP'),
('moved', 'VBD'), ('Monday', 'NNP'), ('to', 'TO'), ('expand', 'VB'),
('background', 'NN'), ('checks', 'NNS'), ('to', 'TO'), ('cover',
'VB'), ('more', 'JJR'), ('firearms', 'NNS'), ('sold', 'VBN'), ('at',
'IN'), ('gun', 'NN'), ('shows', 'NNS'), ('online', 'NN'), ('and',
'CC'), ('anywhere', 'RB'), ('else', 'RB'), ('aiming', 'VBG'), ('to',
'TO'), ('curb', 'VB'), ('a', 'DT'), ('scourge', 'NN'), ('of', 'IN'),
('gun', 'NN'), ('violence', 'NN'), ('despite', 'IN'), ('unyielding',
'JJ'), ('opposition', 'NN'), ('to', 'TO'), ('new', 'JJ'), ('laws',
'NNS'), ('in', 'IN'), ('Congress', 'NNP')]]
```

The word `Obama` was classified as a singular proper noun, `moves` is a verb in the present tense in the singular and third person form, `to` is a preposition but has its own tag, `require` is a verb in its base form, and so on.

The list of part of speech tags (abbreviated) can be found here:
https://www.ling.upenn.edu/courses/Fall_2003/ling001/penn_treebank_pos.html

Knowing the parts of speech of words aids in disambiguating the same sounding words but also helps identify n-grams and named entities. As named entities would normally be spelled starting with a capital letter, we might be tempted to simply list all the NNP and NNPS from our list. Let's see what that brings us:

```
[('House', 'NNP'), ('Philip', 'NNP'), ('Lynch', 'NNP'), ('Department',
'NNP'), ('S', 'NNP'), ('Washington', 'NNP'), ('AP', 'NNP'),
('Constitution', 'NNP'), ('Obama', 'NNP'), ('Duke', 'NNP'),
('University', 'NNP'), ('Monday', 'NNP'), ('Gun', 'NNP'), ('Sen',
'NNP'), ('Tobacco', 'NNP'), ('Firearms', 'NNP'), ('Seattle', 'NNP'),
('Union', 'NNP'), ('State', 'NNP'), ('General', 'NNP'), ('WASHINGTON',
'NNP'), ('Alcohol', 'NNP'), ('Initiative', 'NNP'), ('Attorney',
'NNP'), ('Brady', 'NNP'), ('Iowa', 'NNP'), ('Campaign', 'NNP'),
('Tenn', 'NNP'), ('Republicans', 'NNPS'), ('R', 'NNP'), ('President',
'NNP'), ('Bureau', 'NNP'), ('Cook', 'NNP'), ('Connecticut', 'NNP'),
('Charleston', 'NNP'), ('Loretta', 'NNP'), ('Corker', 'NNP'),
('Republican', 'NNP'), ('Tuesday', 'NNP'), ('C', 'NNP'), ('Dylann',
'NNP'), ('Thursday', 'NNP'), ('Barack', 'NNP'), ('Violence', 'NNP'),
('Explosives', 'NNP'), ('Roof', 'NNP'), ('Times', 'NNP'), ('Clinton',
'NNP'), ('White', 'NNP'), ('Prevent', 'NNP'), ('Americans', 'NNPS'),
('Bob', 'NNP'), ('Justice', 'NNP'), ('Congress', 'NNP'), ('FBI',
'NNP')]
```

It works well enough for some of the words, especially the single ones such as `Republicans` or `Thursday`, but we do not really know whether Philip's last name is Lynch or Cook.

To alleviate this (to some extent), we will use the `.RegexpParser(...)` method offered by NTLK. The method uses regular expressions but looks at part of speech tags instead of the words themselves and matches the patterns:

```
pattern = 'ENT: {<DT>?(<NNP|NNPS>)+}'
```

The pattern that we defined here will tag entities (and flag them as `ENT:`) that follow the structure: any proper noun (for example, `Thursday`) or nouns (for example, `Justice Department`), either singular (for example, `the FBI`) or plural (for example, `Republicans`) that can be preceded by a determiner (for example, `The White House`) or not (for example, `Attorney General Loretta Lynch`).

Extracting the named entities this way yields the following list (you might get these in a different order):

```
[((('a', 'DT'), ('Connecticut', 'NNP')), (('Obama', 'NNP'),), (('the',
'DT'), ('Union', 'NNP')), (('Republicans', 'NNPS'),), (('Corker',
'NNP'),), (('Iowa', 'NNP'),), (('Washington', 'NNP'),), (('Lynch',
'NNP'),), (('State', 'NNP'),), (('the', 'DT'), ('Charleston', 'NNP'),
('S', 'NNP'), ('C', 'NNP')), (('all', 'DT'), ('Americans', 'NNPS')),
(('The', 'DT'), ('Obama', 'NNP')), (('the', 'DT'), ('White', 'NNP'),
('House', 'NNP')), (('a', 'DT'), ('Duke', 'NNP'), ('University',
'NNP')), (('Dylann', 'NNP'), ('Roof', 'NNP')), (('the', 'DT'),
('FBI', 'NNP')), (('Congress', 'NNP'),), (('Tuesday', 'NNP'),),
(('Bureau', 'NNP'),), (('Initiative', 'NNP'),), (('Prevent', 'NNP'),
('Gun', 'NNP'), ('Violence', 'NNP')), (('Philip', 'NNP'), ('Cook',
'NNP')), (('the', 'DT'), ('Republican', 'NNP')), (('Alcohol',
'NNP'), ('Tobacco', 'NNP'), ('Firearms', 'NNP')), (('The', 'DT'),
('Seattle', 'NNP'), ('Times', 'NNP')), (('WASHINGTON', 'NNP'),
('AP', 'NNP'), ('President', 'NNP'), ('Barack', 'NNP'), ('Obama',
'NNP')), (('Monday', 'NNP'),), (('Sen', 'NNP'), ('Bob', 'NNP'),
('Corker', 'NNP'), ('R', 'NNP'), ('Tenn', 'NNP')), (('The', 'DT'),
('White', 'NNP'), ('House', 'NNP')), (('the', 'DT'), ('Brady',
'NNP'), ('Campaign', 'NNP')), (('Explosives', 'NNP'),), (('Thursday',
'NNP'),), (('the', 'DT'), ('Constitution', 'NNP')), (('Gun',
'NNP'),), (('Clinton', 'NNP'),), (('the', 'DT'), ('Justice', 'NNP'),
('Department', 'NNP')), (('Seattle', 'NNP'),), (('Attorney', 'NNP'),
('General', 'NNP'), ('Loretta', 'NNP'), ('Lynch', 'NNP'))]
```

Now we know that Philip's last name was Cook, we properly identified the name of the newspaper The Seattle Times, and properly tagged the *Constitution and Justice Department*. However, we were unable to get the *Bureau of Alcohol, Tobacco, Firearms and Explosives*, and we got a somewhat questionable a *Connecticut* phrase.

Despite these flaws, I think that the named entity identifier worked fairly well and you get a list you can then work with and polish further.

There's more...

Alternatively, we can also use the `.ne_chunk_sents(...)` method provided by NLTK. The beginning of the code is almost identical to the one presented previously (the `nls_pos_alternative.py` file):

```
# extract named entities -- the Named Entity Chunker
ne = [nltk.ne_chunk(ts) for ts in tagged_sentences]

# get a distinct list
named_entities = []
```

```
for s in ne:
    for nent in s:
        if type(nent) == nltk.tree.Tree:
            named_entities.append((nent.label(), tuple(nent)))

named_entities = list(set(named_entities))
named_entities = sorted(named_entities)

# and print out the list
for t, ne in named_entities:
    print(t, ne)
```

The `.ne_chunk_sents(...)` method takes a list of lists of all sentences and words tagged with parts-of-speech and returns a list of sentences with the named entities identified. The named entities are represented as an `nltk.tree.Tree` object.

We then create a distinct list of tuples of all the entities identified and sort it.

 Sorting without specifying the key always sorts given the first element of the tuple.

Finally, we print out the list:

```
FACILITY (('White', 'NNP'), ('House', 'NNP'))
GPE (('Gun', 'NNP'),)
GPE (('Iowa', 'NNP'),)
GPE (('Obama', 'NNP'),)
GPE (('Seattle', 'NNP'),)
GPE (('Washington', 'NNP'),)
GSP (('Connecticut', 'NNP'),)
ORGANIZATION (('AP', 'NNP'),)
ORGANIZATION (('Brady', 'NNP'), ('Campaign', 'NNP'))
ORGANIZATION (('Charleston', 'NNP'),)
ORGANIZATION (('Congress', 'NNP'),)
ORGANIZATION (('Constitution', 'NNP'),)
ORGANIZATION (('Democratic', 'JJ'),)
ORGANIZATION (('Duke', 'NNP'), ('University', 'NNP'))
...
PERSON (('Alcohol', 'NNP'), ('Tobacco', 'NNP'), ('Firearms', 'NNP'))
PERSON (('Attorney', 'NNP'), ('General', 'NNP'), ('Loretta', 'NNP'),
('Lynch', 'NNP'))
PERSON (('Barack', 'NNP'), ('Obama', 'NNP'))
PERSON (('Bob', 'NNP'), ('Corker', 'NNP'))
...
```

The method works well but still could not pick up the more difficult entities such as *Philip Cook* or the *Bureau of Alcohol, Tobacco, Firearms and Explosives*. It also erred by misclassifying Bureau as a person and Obama as GPE.

Identifying the topic of an article

Counting words is a very popular and simple technique that normally renders good results if you want to get a feeling for the topic of the body of text. In this recipe, we will show you how to count the words from The Seattle Times article we have been working with so far to identify the topic of the article without even reading it.

Getting ready

To execute this recipe, you will need NLTK, the regular expressions module from Python, NumPy, and Matplotlib. No other prerequisites are required.

How to do it...

The beginning of the code for this recipe is very similar to the one presented in the previous recipe so we will present only the relevant parts (the nlp_countWords.py file):

```
# part-of-speech tagging
tagged_sentences = [nltk.pos_tag(w) for w in tokenized]

# extract names entities -- regular expressions approach
tagged = []

pattern = '''
    ENT: {<DT>?(<NNP|NNPS>)+}
'''

tokenizer = nltk.RegexpParser(pattern)

for sent in tagged_sentences:
    tagged.append(tokenizer.parse(sent))

# keep named entities together
words = []
lemmatizer = nltk.WordNetLemmatizer()

for sentence in tagged:
    for pos in sentence:
        if type(pos) == nltk.tree.Tree:
```

```
            words.append(' '.join([w[0] for w in pos]))
        else:
            words.append(lemmatizer.lemmatize(pos[0]))

# remove stopwords
stopwords = nltk.corpus.stopwords.words('english')
words = [w for w in words if w.lower() not in stopwords]

# and calculate frequencies
freq = nltk.FreqDist(words)

# sort descending on frequency
f = sorted(freq.items(), key=lambda x: x[1], reverse=True)

# print top words
top_words = [w for w in f if w[1] > 1]
print(top_words)

# plot 10 top words
top_words_transposed = list(zip(*top_words))
y_pos = np.arange(len(top_words_transposed[0][:10]))[::-1]

plt.barh(y_pos, top_words_transposed[1][:10],
    align='center', alpha=0.5)
plt.yticks(y_pos, top_words_transposed[0][:10])
plt.xlabel('Frequency')
plt.ylabel('Top words')

plt.savefig('../../Data/Chapter09/charts/word_frequency.png',
    dpi=300)
```

How it works...

As in the previous recipe, we first import all the necessary modules, read in the data, and do some preprocessing (splitting into sentences and word tokenization).

Next, we tag each word in each sentence with the relevant part of speech and identify the named entities.

For counting purposes, we want to normalize the text a little bit and treat named entities as a single cluster of words. We use `.WordNetLemmatizer()` that we introduced in the *Tokenizing and normalizing text* recipe to normalize the single words. For the named entities, we simply join the list and present it as a string separated by a space, `''.join([w[0] for w in pos]`.

The most common words in English are stopwords, such as `the`, `a`, `and`, or `in` (among many others). Counting such words would give us nothing in understanding the topic of an article or blog post. Therefore, we use the list of stopwords from NLTK to remove all the irrelevant words from our list. The list contains close to 130 words:

```
['i', 'me', 'my', 'myself', 'we', 'our', 'ours', 'ourselves', 'you',
'your', 'yours', 'yourself', 'yourselves', 'he', 'him', 'his',
'himself', 'she', 'her', 'hers', 'herself', 'it', 'its', 'itself',
'they', 'them', 'their', 'theirs', 'themselves', 'what', 'which',
'who', 'whom', 'this', 'that', 'these', 'those', 'am', 'is', 'are',
'was', 'were', 'be', 'been', 'being', 'have', 'has', 'had', 'having',
'do', 'does', 'did', 'doing', 'a', 'an', 'the', 'and', 'but', 'if',
'or', 'because', 'as', 'until', 'while', 'of', 'at', 'by', 'for',
'with', 'about', 'against', 'between', 'into', 'through', 'during',
'before', 'after', 'above', 'below', 'to', 'from', 'up', 'down', 'in',
'out', 'on', 'off', 'over', 'under', 'again', 'further', 'then',
'once', 'here', 'there', 'when', 'where', 'why', 'how', 'all', 'any',
'both', 'each', 'few', 'more', 'most', 'other', 'some', 'such', 'no',
'nor', 'not', 'only', 'own', 'same', 'so', 'than', 'too', 'very', 's',
't', 'can', 'will', 'just', 'don', 'should', 'now']
```

Finally, we use the `.FreqDist(...)` method to count the words. The method takes the list of all the words and simply counts the occurrences. We then sort the list so that we only print out the words that occur more than once; the list contains 83 words and the top five are (not really a big surprise here) `gun`, `Obama`, `background`, `check`, and `said`.

The distribution of the top 10 words looks as follows:

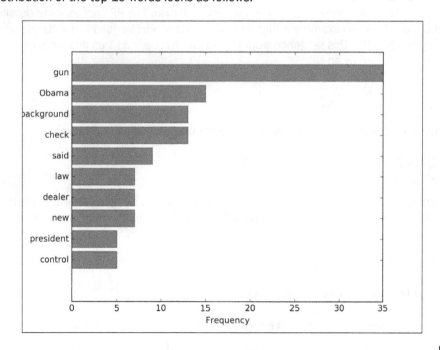

By looking at the distribution of the words, you can infer that the article was about President Obama, gun control, and background checks and somehow involves dealers. We might be a little biased here as we know what the article was about and it is easier for us to pick up the relevant words and put them in context, but as you can see, even with top 10 words you might be able to guess the subject of an article.

Identifying the sentence structure

Another important aspect in understanding free-flow text is the structure of the sentence: we might know the parts of speech but if we do not understand their relationship, we will not understand the greater context of the text.

In this recipe, you will learn how to identify three fundamental parts of any sentence: **noun phrases** (**NP**), **verb phrases** (**VP**), and **prepositional phrases** (**PP**).

The noun phrases always include a noun and its modifiers (or adjectives). An NP can be used to identify a subject of a sentence, although noun phrases can also describe other objects in the sentence. For example, the sentence, `My dog is lying on the carpet`, has two NPs: `my dog` and `on the carpet`; the first is the subject and the latter NP is the object.

The verb phrases always include at least one verb and objects, complements, and its modifiers. It normally describes an action that the subject exerts on the object. Considering the previous sentence, the VP would be `is lying on the carpet`.

The prepositional phrases will almost always start with a preposition (`at`, `in`, `from`, `with`, and so on) and end with a noun, pronoun, gerund, or clause. The PPs act as more detailed adverbs or adjectives. An example would be `A dog with white tail is laying on the carpet`; the PP in this sentence would be `with white tail` as it is a combination of a preposition followed by an NP.

Getting ready

To execute this recipe, you will need `NLTK`. No other prerequisites are required.

How to do it...

In this recipe, to present the mechanism, we only selected two sample sentences from The Seattle Times article—one relatively simple and another slightly more complex (the `nlp_sentence.py` file):

```
import nltk

def print_tree(tree, filename):
    '''
        A method to save the parsed NLTK tree to a PS file
```

```
'''
    # create the canvas
    canvasFrame = nltk.draw.util.CanvasFrame()

    # create tree widget
    widget = nltk.draw.TreeWidget(canvasFrame.canvas(), tree)

    # add the widget to canvas
    canvasFrame.add_widget(widget, 10, 10)

    # save the file
    canvasFrame.print_to_file(filename)

    # release the object
    canvasFrame.destroy()

# two sentences from the article
sentences = ['Washington state voters last fall passed Initiative
594', 'The White House also said it planned to ask Congress for $500
million to improve mental health care, and Obama issued a memorandum
directing federal agencies to conduct or sponsor research into smart
gun technology that reduces the risk of accidental gun discharges.']

# the simplest possible word tokenizer
sentences = [s.split() for s in sentences]

# part-of-speech tagging
sentences = [nltk.pos_tag(s) for s in sentences]

# pattern for recognizing structures of the sentence
pattern = '''
    NP: {<DT|JJ|NN.*|CD>+}    # Chunk sequences of DT, JJ, NN
    VP: {<VB.*><NP|PP>+}      # Chunk verbs and their arguments
    PP: {<IN><NP>}            # Chunk prepositions followed by NP
  '''

# identify the chunks
NPChunker = nltk.RegexpParser(pattern)
chunks = [NPChunker.parse(s) for s in sentences]

# save to file
print_tree(chunks[0], '../../Data/Chapter09/charts/sent1.ps')
print_tree(chunks[1], '../../Data/Chapter09/charts/sent2.ps')
```

How it works...

We store the two sentences in a list. As these are fairly simple, we use the simplest word tokenizer possible; we split the sentence at each space character. This is followed by the already familiar part-of-speech tagging.

We have already used `.RegexpParser(...)` earlier but with a different pattern.

As outlined in the introduction to this recipe, the NP consists of a noun and its modifiers. The pattern here states exactly that an NP is a combination of one or more DTs (determiners, such as `the` or `a`), JJs (adjectives), CDs (cardinal numbers), or any variation of a noun: NN.* would unfold to NN (noun, singular), NNS (noun, plural), NNP (proper noun, singular), and NNP (proper noun, plural).

The VP is a combination of one or more verbs of any form: VB (base form), VBD (past tense), VBG (gerund or present participle), VBN (past participle), VBP (non-third person singular present), VBZ (third person singular present), and either NP or PP.

The PP, as explained earlier, is a combination of a preposition (IN) and NP.

Once parsed, the sentence trees can be saved in a file. We use the `print_tree(...)` method passing `tree` as the first argument and `filename` as the second one.

In the method, we first create `canvasFrame` and add a tree widget to it. The `.TreeWidget(...)` method takes the newly created canvas as its first argument and the `tree` as its second argument. Next, we add widget to `canvasFrame` and save it in a file. Finally, we release the memory occupied by `canvasFrame`.

However, the only supported format to save the tree is postscript. To convert the postscript file to a PDF format, you can use one of many available online converters, for example, `https://online2pdf.com/convert-ps-to-pdf` or, on a Unix-like machine, you can issue the following command (assuming that you are in the `Codes/Chapter09` folder):

```
convert -density 300 ../../Data/Chapter09/charts/sent1.ps ../../Data/
Chapter9/charts/sent1.png
```

The whole command should be typed in one line (the formatting of the book prevents us from doing so here).

The resulting tree (for the first sentence) looks as follows:

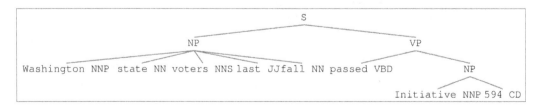

We have our NP that contains the subject: `Washington state voters last fall`, the verb: `passed`, and the object (another NP): *Initiative NNP 594*.

See also

Here's another example on identifying the structural parts of a sentence:

`https://www.eecis.udel.edu/~trnka/CISC889-11S/lectures/dongqing-chunking.pdf`

Classifying movies based on their reviews

Having gone through all the preparations, we are now ready to embark on a more advanced task: classifying movies based on their reviews. In this recipe, we will use a sentiment analyzer and Naïve Bayes classifier to classify the movies.

Getting ready

To execute this recipe, you will need `NLTK` and `JSON`. No other prerequisites are required.

How to do it...

It takes a bit of wrangling but ultimately the code is fairly easy to understand and follow (the `nlp_classify.py` file):

```python
# this is needed to load helper from the parent folder
import sys
sys.path.append('..')

# the rest of the imports
import helper as hlp
import nltk
import nltk.sentiment as sent
import json


@hlp.timeit
def classify_movies(train, sentim_analyzer):
    '''
        Method to estimate a Naive Bayes classifier
        to classify movies based on their reviews
    '''
    nb_classifier = nltk.classify.NaiveBayesClassifier.train
```

```
            classifier = sentim_analyzer.train(nb_classifier, train)

            return classifier

    @hlp.timeit
    def evaluate_classifier(test, sentim_analyzer):
        '''
            Method to estimate a Naive Bayes classifier
            to classify movies based on their reviews
        '''
        for key, value in sorted(sentim_analyzer.evaluate(test).items()):
            print('{0}: {1}'.format(key, value))

    # read in the files
    f_training = '../../Data/Chapter09/movie_reviews_train.json'
    f_testing  = '../../Data/Chapter09/movie_reviews_test.json'

    with open(f_training, 'r') as f:
        read = f.read()
        train = json.loads(read)

    with open(f_testing, 'r') as f:
        read = f.read()
        test = json.loads(read)

    # tokenize the words
    tokenizer = nltk.tokenize.TreebankWordTokenizer()

    train = [(tokenizer.tokenize(r['review']), r['sentiment'])
        for r in train]

    test  = [(tokenizer.tokenize(r['review']), r['sentiment'])
        for r in test]

    # analyze the sentiment of reviews
    sentim_analyzer = sent.SentimentAnalyzer()
    all_words_neg_flagged = sentim_analyzer.all_words(
        [sent.util.mark_negation(doc) for doc in train])

    # get most frequent words
    unigram_feats = sentim_analyzer.unigram_word_feats(
        all_words_neg_flagged, min_freq=4)

    # add feature extractor
```

```
sentim_analyzer.add_feat_extractor(
    sent.util.extract_unigram_feats, unigrams=unigram_feats)

# and create the training and testing using the newly created
# features
train = sentim_analyzer.apply_features(train)
test  = sentim_analyzer.apply_features(test)

# what is left is to classify the movies and then evaluate
# the performance of the classifier
classify_movies(train, sentim_analyzer)
evaluate_classifier(test, sentim_analyzer)
```

How it works...

As always, we start with importing all the necessary modules.

The training and testing files that we prepared earlier are stored in the `Data/Chapter09` folder in a JSON format. You should know how to read JSON formatted files; if you need to refresh, refer to the *Reading and writing JSON files with Python* recipe from *Chapter 1, Preparing the Data*.

Having read the files, in this recipe, we use `.TreebankWordTokenizer()` to tokenize our reviews. Note that we do not even split the reviews into sentences; the idea for this recipe is to identify words in each review that have a negative meaning to what they would normally represent and, based on such features, attempt to classify whether the review was positive or negative.

Once tokenized, the `train` and `test` objects are lists of tuples with the first element being the tokenized list of words of the review and the second element identifying the sentiment of the review: either `pos` or `neg`.

Once we have our dataset prepared, we can create the feature set that we will use to train our Naïve Bayes classifier. We use `.SentimentAnalyzer()` from NLTK.

 Check the documentation for the sentiment analysis from NLTK:
`http://www.nltk.org/api/nltk.sentiment.html`

The `.SentimentAnalyzer()` helps streamline the tasks of feature extraction and classification.

First, we flag all the words that have a changed meaning because they were preceded by a negation. Consider the following sentences: `I love going out. I didn't like yesterday's outing, though.` Obviously, the word `like` in this context isn't really positive as it is preceded by `didn't`. The `.util.mark_negation(...)` method takes a list of words and marks those followed by negation with a suffix, `_NEG`; in our preceding example, we would have gotten back a list as follows:

```
[['I', 'love', 'going', 'out']['I', "didn't", 'like_NEG',
"yesterday's_NEG", 'outing_NEG', 'though_NEG']]
```

Now that we have all the words flagged, we can create unigram features using these. In our example, we want to use only the words that appeared above a certain level: seeing the word `like_NEG` only once would not make a good predictor. We use the `.unigram_word_feats(...)` method that returns the most common words (if we specify the `top_n` parameter) or all the words with a frequency greater than `min_freq`.

The `.add_feat_extractor(...)` method adds a new method to the `SentimentAnalyzer()` object to extract words from documents. We will be only focusing on unigrams in this example, so we pass the `.util.extract_unigram_feats` method as the first argument to the `.add_feat_extractor(...)` method. The unigrams keyword parameter is the list of our unigram features.

Once we have the feature extractor ready, we can transform our training and testing datasets. We use the `.apply_features(...)` method that transforms our list of words into a list of feature vectors, each element of which indicates whether the review contains a specific word/feature or not. If you were to print it out, it would look as follows (abbreviated for obvious reasons):

```
..., 'contains(cinematic)': False, 'contains(nurse_NEG)': False},
'pos'),
```

What is left is to train the classifier and later evaluate it. We pass the training dataset and the `sentiment_analyzer` object to the `classify_movies(...)` method. The `.train(...)` method can accept various trainers but here we will use the `.classify.NaiveBayesClassifier.train` method. We also pass the training dataset. Using the `.timeit` decorator, we measured the time it took to estimate the model on our machine:

```
Training classifier
The method classify_movies took 552.16 sec to run.
```

So, it almost took over nine minutes to train the classifier. What is more important, though, is how well it performed—remember that Naïve Bayes is one of the simpler classification methods (refer to the *Classifying with Naïve Bayes* recipe from *Chapter 3, Classification Techniques*).

The `evaluate_classifier(...)` method takes the testing dataset and trained classifier, runs the evaluation, and prints out the results. The overall accuracy of the built classifier is not too bad, close to 82%. This shows that our training dataset generalized the problem well:

```
Evaluating NaiveBayesClassifier results...
Accuracy: 0.817
F-measure [neg]: 0.8211143695014662
F-measure [pos]: 0.812691914022518
Precision [neg]: 0.8030592734225621
Precision [pos]: 0.8322851153039832
Recall [neg]: 0.84
Recall [pos]: 0.794
The method evaluate_classifier took 1954.13 sec to run.
```

The values of `Precision` and `Recall` are really good for a fairly simple classifier. It took, however, over half an hour to evaluate the model.

10
Discrete Choice Models

In this chapter, you will learn the following recipes:

- ▶ Preparing a dataset to estimate discrete choice models
- ▶ Estimating the well-known Multinomial Logit model
- ▶ Testing for violations of the Independence from Irrelevant Alternatives
- ▶ Handling IIA violations with the Nested Logit model
- ▶ Managing sophisticated substitution patterns with the Mixed Logit model

Introduction

Discrete choice models (**DCMs**) aim at predicting which alternative a person will choose. The models share similarities with logistic regression although with some fundamental differences in assumptions about the distribution of error terms.

The theory of DCMs has its roots in the random utility theory and assumption that a rational person will always choose an option that will maximize the utility a person gets from choosing such an option.

For example, let's assume that you are to choose between two independent alternatives (that is, such alternatives that do not share any common characteristics); for the sake of this example, we will consider choosing between biking and driving a car to work. It costs nothing to bike to work (we are, of course, assuming that you already have a bike and we do not count the energy that you burn while biking), but the cost of driving a car to work would be $3. However, it takes roughly 45 minutes to get to work on a bike while the same trip using a car would be only 8 minutes. Assuming -2 for a price coefficient and -0.13 for the travel time, we can calculate the utility of riding a bike to be -5.85 while driving a car would be -7.04. Therefore, under the utility maximization assumption, you would be more likely to choose to ride a bike than take a car to work. Our simplified calculations do not take into account the characteristics of the person.

For the purpose of the recipes in this chapter, we generated a dataset of 10,000 choices among 16 different booking classes on four different flights (between Seattle and Los Angeles); all the options are summarized in the following table:

Flight	Booking class	Price	Compartment	Frequent flyer points	Fully refundable
AA777	C	$410	business	yes	yes
	Z	$320	business	yes	no
	Y	$210	economy	yes	yes
	V	$150	economy	no	no
AS666	C	$390	business	yes	yes
	Z	$315	business	yes	no
	Y	$195	economy	yes	yes
	V	$130	economy	no	no
DL001	C	$400	business	yes	yes
	Z	$320	business	yes	no
	Y	$200	economy	yes	yes
	V	$130	economy	no	no
UA110	C	$380	business	yes	yes
	Z	$310	business	no	no
	Y	$200	economy	yes	yes
	V	$125	economy	no	no

In the next chapter, we will see how to transform this dataset so that we can use it to estimate various DCMs.

For the purpose of estimating DCMs, we will be using Python Biogeme, introduced by Dr. Michel Bierlaire in *Bierlaire, M. (2003); BIOGEME: A free package for the estimation of discrete choice models*, Proceedings of the *3rd Swiss Transportation Research Conference, Ascona, Switzerland*. Although not necessarily pure Python (as the estimation engine is written in C++), it is the best open source package to estimate DCMs. The Python Biogeme is actively being developed and can be downloaded from `http://biogeme.epfl.ch`; just go to the **Download** section and choose the version for your operating system. As the package is distributed free of charge, you need to register to Biogeme's user group on Yahoo via `https://groups.yahoo.com/neo/groups/biogeme/info`.

Once you download the package, follow the instructions outlined at `http://biogeme.epfl.ch/install.html` to install the package. Once installed, you can invoke the package from the command line in the following form:

```
pythonbiogeme <model_file> <dataset>
```

Cautionary note

We have had quite a few problems getting the package to work. The package requires Python 3.2 or a more recent version. In our case, the installer installed the package in `/usr/local/bin`. However, even though we use Python 3.5 in this book (located in the `~/anaconda` folder), Python Biogeme was searching for the Python installation in `/usr/local/Frameworks/Python.framework/Versions/3.5`—a folder that did not exist on our machine. We manually created the folder and then installed the package from the image again.

We also tried running the package on Windows and it ran properly. However, it was brought to our attention that some users might have problems with installing and running the package.

Should you encounter any problems, problems you can ask more specific questions to the Biogeme's user group; he usually responds to queries fairly quickly. If you are truly desperate, you can also send me an e-mail at `mail@tomdrabas.com` and if I am able to help, I will.

Preparing a dataset to estimate discrete choice models

The dataset prepared for these recipes consists of 10,000 choice situations. The first column contains the ID of the chosen alternative and the second column contains a list of all the alternatives the person was choosing from; not all of the alternatives were available in all choice situations.

Getting ready

To execute this query, you will need `pandas`, `NumPy`, and the regular expressions module. No other prerequisites are required.

How to do it...

Python Biogeme requires a tab-separated dataset, where each row contains the attributes of all the alternatives and the flag for the chosen one (an integer) as well as an indicator whether the alternative was available to the decision-maker at the time when the decision was made. Only numerical values are allowed. The following code can be found in the `dcm_dataPrep.py` file:

```python
import pandas as pd
import numpy as np
import re

# read datasets
observations_filename = '../../Data/Chapter10/observations.csv'
alternatives_filename = '../../Data/Chapter10/options.json'

observations = pd.read_csv(observations_filename)
alternatives = pd.read_json(alternatives_filename).transpose()

# retrieve all considered alternatives
considered = [alt.split(';')
    for alt in list(observations['available'])]

# create flag of all available alternatives
available = []
alternatives_list = list(alternatives.index)
alternatives_list = [
    re.sub(r'\.', r'_', alt) for alt in alternatives_list]
no_of_alternatives = len(alternatives_list)

for cons in considered:
    f = [0] * no_of_alternatives
    cons = [re.sub(r'\.', r'_', alt) for alt in cons]

    for i, alt in enumerate(alternatives_list):
        if alt in cons:
            f[i] = 1

    available.append(list(f))

# append to the choices DataFrame
alternatives_av = [alt + '_AV' for alt in alternatives_list]
available = pd.DataFrame(available, columns=alternatives_av)
observations = observations.join(available)
```

```
# drop the available column as we don't need it anymore
del observations['available']

# encode the choice variable
observation = list(observations['choice'])
observation = [re.sub(r'\.', r'_', alt) for alt in observation]
observation = [alternatives_list.index(c) + 1 \
    for c in observation]

observations['choice'] = pd.Series(observation)

# and add the alternatives' attributes
# first, normalize price to be between 0 and 1
max_price = np.max(alternatives['price'])
alternatives['price'] = alternatives['price'] / max_price

# next, create a vector with all attributes
attributes = []
attributes_list = list(alternatives.values)

for attribute in attributes_list:
    attributes += list(attribute)

# fill in to match the number of rows
attributes = [attributes] * len(choices)

# and their names
attributes_names = []

for alternative in alternatives_list:
    for attribute in alternatives.columns:
        attributes_names.append(alternative + '_' + attribute)

# convert to a DataFrame
attributes = pd.DataFrame(attributes,
    columns=attributes_names)

# and join with the main dataset
observations = choices.join(attributes)

# save as a TSV with .dat extension
with open('../../Data/Chapter10/observations.dat', 'w') as f:
    f.write(observations.to_csv(sep='\t', index=False))
```

How it works...

As always, we first load all the modules that we will use.

The dataset we generated is located in the `Data/Chapter10/` folder in the Git repository for the book. It is split into two files: `observations.csv` that contains the 10,000 observations with choices made, and `options.json` that contains a list of all the alternatives with their attributes.

The `observations.csv` file has the following format:

```
choice,available
AA777.4.V,AA777.1.C;AA777.2.Z;AA777.4.V;AS666.1.C;AS666.3.Y;AS666.4.V;
DL001.2.Z;DL001.3.Y;DL001.4.V;UA110.1.C;UA110.3.Y;UA110.4.V
. . .
```

The first column contains the option chosen: in this example, it was booking class V on the AA777 flight. The person chose this one from 12 other booking classes: the list that is separated by `;` lists 12 available options.

The `options.json` file has the following format:

```
{
    "AA777.1.C": {
        "compartment": 1,
        "frequentFlyer": 1,
        "price": 410,
        "refund": 1
    },
    "AA777.2.Z": {
. . .
```

The `compartment`, when set to 1, indicates business class, and 1s for `frequentFlyer` and `refund` indicate that while buying such a booking class, a person would get frequent flyer points and can return the ticket without penalty.

 To read CSV- or JSON-formatted data using `pandas`, refer to the *Reading and writing CSV/TSV files with Python* recipe in *Chapter 1, Preparing the Data*.

Having read the datasets, we turn our attention to the list of considered alternatives first. From the `choices` DataFrame, we extract the available column and split the rows into individual alternatives. We will use this to create a DataFrame with all the available (considered) alternatives.

In order to do so, we need a list of all the available alternatives. We get that by extracting the index from the `alternatives` DataFrame. However, the alternative indicators contain full stops, for example, `AA777.1.C`. This will cause problems when we move to Python Biogeme, so we use the `.sub(...)` method of `re` package to substitute the full stops with underscores.

> Note that a regular expression `r'.'` would effectively substitute all the characters in any string with the underscore as `r'.'` matches any character but a newline. We need to use the escape character `r'\.'` to match only full stops.

The `.sub(...)` method takes the pattern to be found in the string as its first parameter, the replacement string as the second parameter, and the string to replace the patterns in as the third one.

Finally, we loop through all the records and mark the available alternatives. First, we create a temporary list with as many `0`s as we have alternatives. Then, in order to keep it compatible with our `alternatives_list`, we also replace all the full stops in the considered alternatives cons. Finally, we loop through all the alternatives and mark those available with `1`s. We would end up with the following list (for each choice situation):

```
[1, 1, 0, 1, 1, 0, 1, 1, 0, 1, 1, 1, 1, 0, 1, 1]
```

Let's create the names for the corresponding columns so that we can use these flags later in Python Biogeme; to accomplish this, we simply append the `_AV` suffix to each alternative's name.

Finally, we create a DataFrame `available` that contains all the availability vectors, join it with the `choices` DataFrame, and drop the `available` column as we do not need it anymore:

	choice	AA777_1_C_AV	AA777_2_Z_AV	AA777_3_Y_AV	AA777_4_V_AV \
0	AA777.4.V	1	1	0	1
1	UA110.3.Y	1	1	1	1
2	DL001.1.C	1	1	1	1
3	AS666.4.V	1	1	1	1
4	DL001.2.Z	1	1	1	1

	AS666_1_C_AV	AS666_2_Z_AV	AS666_3_Y_AV	AS666_4_V_AV	DL001_1_C_AV \
0	1	0	1	1	0
1	1	1	0	1	0
2	0	0	1	1	1
3	1	0	1	1	1
4	1	1	1	1	1

	DL001_2_Z_AV	DL001_3_Y_AV	DL001_4_V_AV	UA110_1_C_AV	UA110_2_Z_AV \
0	1	1	1	1	0
1	1	0	0	0	1
2	0	1	0	0	0
3	1	1	1	1	1
4	1	1	1	1	1

	UA110_3_Y_AV	UA110_4_V_AV
0	1	1
1	1	1
2	1	1
3	1	1
4	1	1

As you can see, the booking class V on flight AA777 was chosen and, as expected, it was considered by the decision-maker.

Next, we need to encode the alternatives in the `choice` column. First, we retrieve all the records from the DataFrame and substitute the full stops with underscores. Next, we simply loop through all the elements and find `.index(...)` of that element in `alternatives_list`. By doing so, we keep it consistent with the rest of the dataset. Finally, we substitute the `choice` column with the newly encoded `choice` list that we cast into a `Series` object of pandas.

The last thing that we need to take care of is adding all the attributes of all the alternatives to our dataset. As I mentioned earlier, each row in the dataset needs to contain all the information about all the alternatives.

In our case, it is a list of essentially the same information, top to bottom. However, this approach is much more useful if you deal with more generic alternatives, for example, train versus car choice, as then the price and distance for each row can be different. If our dataset had different destinations and we were modeling a choice of airline and booking class (not flight), each row could have different attributes.

First, we normalize the price variable to be between 0 and 1. We simply retrieve the maximum price from the dataset and divide all the prices by it. We could achieve the same in Python Biogeme but I prefer doing it upfront and save Python Biogeme just to estimate the models.

Next, we retrieve all the values for the attributes from the `alternatives` DataFrame by extracting the `.values` attribute of the DataFrame. The `attributes_list` is essentially a list of `NumPy` arrays (as pandas' underlying data structures borrow heavily from `NumPy`):

```
[array([ 1.,    1.,    1.,    1.]), array([ 1.        ,    1.        ,
 0.7804878,  0.        ]), ...]
```

We need a long vector with all the attributes so we loop through `attributes_list` and extend the `attributes` list by each array. We need to copy it for all 10,000 rows, so we multiply the `attributes` list by the length of `choices`, that is, the number of rows.

Finally, we create the name of the columns for all the attributes. Once we have the `attributes` list created and the names of the columns in place, we create a DataFrame and join it with the original dataset.

As the last step, we export the dataset to a tab-separated `.dat` file (to keep it consistent with Biogeme's tradition).

There's more...

If you have a full dataset in a CSV format, you can use the `bioprepdata` tool of Biogeme to create the dataset. Also, `biocheckdata` can be used to check whether the dataset complies with Python Biogeme's standards.

Estimating the well-known Multinomial Logit model

The **Multinomial Logit** (**MNL**) model was introduced by Daniel McFadden in his seminal paper from 1973, `http://eml.berkeley.edu/reprints/mcfadden/zarembka.pdf`.

The model is based on fairly restrictive assumptions: the **Independent and Identically Distributed** (**IID**) error terms and the **Independence from Irrelevant Alternatives** (**IIA**).

The IID assumes that the error terms of utility functions of all the alternatives are independent (uncorrelated) and follow the same distribution. (For MNL, it is the Extreme Value Type I Distribution, commonly known as Gumbel distribution after E. J. Gumbel who derived and analyzed it.)

The IIA, on the other hand, assumes that the ratios of probabilities between alternatives are constant, that is, removing one or more alternatives from the consideration set does not change the ratio between the remaining alternatives. Consider the following situation: you are choosing between a bike, train, and car to get to work. For the sake of this example, let's assume that each of the alternatives is equally probable, that is, each has a probability of being chosen equal to 1/3. Under the IIA, if your car broke, the ratio or probabilities between the other two alternatives would remain constant, that is, each would have a 50/50 chance of being chosen. Why this is restrictive we will discuss in the next recipe.

The probabilities in MNL are calculated as follows (for alternative `i`):

$P(i) = exp(U(i)) / \Sigma_j \, exp(U(j))$

Here, $U(i)$ is the utility of each alternative defined as $U(i) = \alpha_i + \beta^i X_i + \varepsilon_i$ where α_i is the alternative specific constant, β^i is a vector of alternative i specific coefficients of attributes, X_i is a vector of attributes for alternative i, and ε_i are the error terms.

Getting ready

To execute this recipe, you need a working Python Biogeme package installed on your machine. No other prerequisites are required.

How to do it...

The syntax used in Python Biogeme is fairly similar to Python itself. Python, for Biogeme, is essentially a thin wrapper around the underlying C++ code (the `dcm_mln.py` file in the MNL folder). The following code is in abbreviated form to save space:

```
from biogeme import *
from headers import *
from loglikelihood import *
from statistics import *

# Specify parameters to be estimated
ASC     = Beta('ASC',0,-10,10,1,'ASC')
C_price = Beta('C_price',0,-10,10,0,'C price')
Z_price = Beta('Z_price',0,-10,10,0,'Z price')
Y_price = Beta('Y_price',0,-10,10,0,'Y price')
V_price = Beta('V_price',0,-10,10,0,'V price')

B_comp   = Beta('B_comp',0,-10,10,0,'compartment')
B_refund = Beta('B_refund',0,-3,3,0,'refund')

# Utility functions

# The data are associated with the alternative index
# compartment attributes
c = {}

c[1]  = AA777_1_C_compartment
...
c[16] = UA110_4_V_compartment

# price attributes
p = {}

p[1]  = AA777_1_C_price
...
p[16] = UA110_4_V_price

# refund attributes
r = {}

r[1]  = AA777_1_C_refund
...
```

```
r[16] = UA110_4_V_refund

# The dictionary of utilities is constructed.
V = {}

V[1] = C_price * p[1] + B_refund * r[1] + B_comp * c[1]
V[2] = Z_price * p[2] + B_refund * r[2] + B_comp * c[2] + ASC
...
V[16] = Y_price * p[16] + B_refund * r[16] + B_comp * c[16]

# availability flags
availability = {
    1:  AA777_1_C_AV,
    ...
    16: UA110_4_V_AV
}

# The choice model is a logit, with availability conditions
logprob = bioLogLogit(V, availability, choice)

# Defines an itertor on the data
rowIterator('obsIter')

# Define the likelihood function for the estimation
BIOGEME_OBJECT.ESTIMATE = Sum(logprob, 'obsIter')

# Statistics
nullLoglikelihood(availability,'obsIter')
choiceSet = [1,2,3,4,5,6,7,8,9,10,11,12,13,14,15,16]
cteLoglikelihood(choiceSet, choice, 'obsIter')
availabilityStatistics(availability, 'obsIter')

# Parameters
BIOGEME_OBJECT.PARAMETERS['optimizationAlgorithm'] = 'BIO'
BIOGEME_OBJECT.PARAMETERS['numberOfThreads'] = '8'

BIOGEME_OBJECT.FORMULAS['AA777 C utility'] = V[1]
...
BIOGEME_OBJECT.FORMULAS['UA110 V utility'] = V[16]
```

How it works...

To execute this code, you can either use the `run_pythonbiogeme.sh` shell script in the main folder for the chapter or run the following (assuming that you are in the MNL folder):

```
pythonbiogeme dcm_mnl ../../../Data/Chapter10/choices.dat
```

 The `run_pythonbiogeme.sh` removes files created by previous runs of `pythonbiogeme` before estimating the model again.

Using the `run_pythonbiogeme.sh` script is also very simple (the whole command should be typed in the same line):

```
../run_pythonbiogeme.sh dcm_mnl ../../../Data/Chapter10/choices.dat
```

Using either of the commands, you will see something as follows:

```
This is biogeme (pythonbiogeme) 2.4
 Read sample file: ../../../Data/Chapter10/choices.dat
 Nbr of cores used by biogeme: 8
 Init. log-likelihood: -25531.5 [00:00]
     gmax Iter    radius         f(x)      Status       rhok nFree
 +3.92e-04    1 1.00e+00 +2.5531498e+04 ****OutTrReg +9.96e-01 6   ++ P
 +4.47e-04    2 2.00e+00 +2.4122563e+04 ****OutTrReg +9.86e-01 6   ++ P
 +1.24e-03    3 4.00e+00 +2.2093986e+04 ****Converg +1.09e+00 5   ++ P
 +1.25e-03    4 8.00e+00 +2.1652130e+04 ****Converg +1.06e+00 5   ++ P
 +8.32e-04    5 1.60e+01 +2.1635928e+04 ****Converg +1.02e+00 6   ++ P
 +3.96e-04    6 3.20e+01 +2.1617511e+04 ****Converg +1.01e+00 6   ++ P
 +3.10e-05    7 6.40e+01 +2.1617465e+04 ****Converg +1.00e+00 6   ++ P


 Convergence reached...
--> time interval [22:59:21,22:59:22]
 Estimated parameters:
C_price = -3.13244
B_refund = -0.617234
B_comp = -0.673158
Y_price = -4.91588
Z_price = -3.44861
ASC = 0
```

```
V_price = -5.52783
  File dcm_mnl_param.py created
  File dcm_mnl.html has been generated
  File dcm_mnl.tex has been generated
```
Done

`Init. Log-likelihood` gives the initial value of the `log-likelihood` function, that is, the value of the `log-likelihood` function with initial values of coefficients. This, effectively, means that each alternative is equally probable to be selected. This value is later used to calculate the rho-squared value, an equivalent of R^2 from Linear Regression. The aim of estimating any DCM is to minimize the absolute value of the `log-likelihood` function. Next, the output table shows the progression of the estimation; you can see that the value of `f(x)` (our `log-likelihood`) gets smaller and smaller (as expected) until iteration 7 when the iterations stop, as the difference between the value of the `log-likelihood` function between iterations 6 and 7 is small enough and so the function has converged. The values of estimated parameters are printed out below the table but we will look at them in the `dcm_mnl.html` file later.

Let's now go through the model file step by step.

First, as always, we load the necessary modules. The `biogeme`, `headers`, `log-likelihood`, and `statistics` are the necessary modules that almost all of your models will require.

Next, we specify the coefficients using the `Beta(...)` method. The method's first argument is the name of the parameter (normally, the same as the name of the object). The second parameter defines the initial value of the coefficient. The third and fourth parameters determine the lower and higher bounds of the value of the coefficient, and the fifth parameter determines whether the value of the coefficient is to be fixed (not estimated) or estimated.

For the model to be identifiable, one of the parameters needs to be normalized, that is, fixed to 0. Normally, it is one of the **alternative specific constants** (**ASCs**). Check *Chapter 2, Properties of Discrete Choice Models* of K. Train's book, `http://eml.berkeley.edu/books/choice2nd/Ch02_p9-33.pdf`

The last parameter for the `Beta(...)` method is a friendly name that will be used in the parameters' report.

Having specified the coefficients, we create dictionaries of attribute names.

These names must appear exactly as stated in your `.dat` file.

We will only be using the `compartment`, `price`, and `refund` attributes of each alternative when we specify the utility functions for each alternative. Note that, to avoid confusion, we store the attributes of the same alternatives under the same number in our dictionaries.

Finally, the `V` dictionary holds all of our utility functions. We include `ASC` only for alternative 2 (the flight `AA777`, booking class `Z`). Note also that the `B_refund` and `B_comp` coefficients are not alternative-specific but generic—the same for all the alternatives. In contrast, the price coefficients are shared by the equivalent booking classes across flights.

The `availability` object specifies the availability flags for each of the alternatives.

Finally, the estimation parameters start with specifying the `logprob` object. To estimate the MNL, we use the `bioLogLogit(...)` method. The method takes the dictionary of utilities as its first parameter, the availability dictionary as its second parameter, and the name of the variable that holds `choice`. Next, we define an iterator over all the observations in our dataset and call it **obsIter**. Finally, for `BIOGEME_OBJECT`, we specify the `ESTIMATE` variable: this specifies that the sum of log probabilities of the whole dataset is to be minimized.

The remainder of the model specification deals with calculating statistics of the estimates. The `nullLogLikelihood(...)` method calculates the null `log-likelihood` and includes it in the output. The first parameter is the dictionary with the availability of each of the alternatives and the other one is the name of the row iterator. The `cteLoglikelihood(...)` calculates the value of the `log-likelihood` function with only constant parameters included. It takes `choiceSet` (a list of all the alternatives) as its first argument, the name of the variable that holds the choice outcomes as the second, and the iterator over the records as the last one. The final statistic that we will be including is `availabilityStatistics(...)` that returns the count of time each alternative is available in the dataset.

Lastly, we specify `BIOGEME_OBJECT.PARAMETERS` and `.FORMULAS`. With the `.PARAMETERS(...)` method, you can pass various parameters to both Biogeme and its solvers; here, we specify the `BIO` optimizer (which is Biogeme's own optimization algorithm) and the number of threads to start when executing. The `.FORMULAS(...)` method allows us to give a friendly name to our utility functions and also print them out in the final report.

The Python Biogeme report looks as follows (abbreviated; for the full version, open the dcm_mn1.html file in your browser):

Estimation report

Number of estimated parameters: 6
Sample size: 10000
Excluded observations: 0
Init log likelihood: -25531.498
Final log likelihood: -21614.578
Likelihood ratio test for the init. model: 7833.839
Rho-square for the init. model: 0.153
Rho-square-bar for the init. model: 0.153
Final gradient norm: +2.633e-03
Diagnostic: Convergence reached...
Iterations: 7
Run time: 00:01
Nbr of threads: 8

Estimated parameters

Click on the headers of the columns to sort the table [Credits]

Name	Value	Std err	t-test	p-value	Robust Std err	Robust t-test	p-value
B_comp	3.53	1.30	2.70	0.01	1.31	2.70	0.01
B_refund	-0.719	0.137	-5.24	0.00	0.137	-5.23	0.00
C_price	-7.30	1.33	-5.50	0.00	1.33	-5.49	0.00
V_price	-5.07	0.648	-7.83	0.00	0.647	-7.84	0.00
Y_price	-4.41	0.708	-6.23	0.00	0.706	-6.24	0.00
Z_price	-8.71	1.65	-5.27	0.00	1.66	-5.25	0.00

The report specifies a multitude of various statistics. Some of them we have covered already, so here we will only focus on those that are of more interest. You should be looking at the Rho-square-bar for the init. model statistic: this statistic tells you how good your model is. Values above 0.20 are considered very good (equivalent of R^2 of 85% for a linear model), between 0.15 and 0.2 are good, values of 0.1 to 0.15 are acceptable, and a value of this statistic below 0.1 depicts a poor model.

Importantly, Diagnostic should always read Convergence reached. As we will see later, the Run time is another important statistic to look at, especially when you are dealing with more sophisticated models.

As far as the Estimated parameters are concerned, you should be keeping all the variables that are statistically significant, that is, with p-values < 0.05. In our model, all of them are statistically significant. Also, all the price coefficients are negative, meaning that a higher price would reduce the chances of the alternative to be selected. Surprisingly, however, if an option is offered in a business class, its chances of being selected increase.

See also

There are numerous places to read about MNL:

- ▶ A great (free) book by Kenneth Train can be found here:
  `http://eml.berkeley.edu/books/choice2.html`

- ▶ You can also check out this presentation:
  `http://www.bauer.uh.edu/rsusmel/phd/ec1-20.pdf`

- ▶ This is the introduction on Python Biogeme:
  `http://biogeme.epfl.ch/documentation/pythonfirstmodel-2.4.pdf`

Testing for violations of the Independence from Irrelevant Alternatives

The MNL is based on a fairly restrictive IIA property that assumes that the ratios of probabilities of the alternatives remain unchanged. This is true only for the choice set (set of all the alternatives) that does not share any common characteristic or, put differently, the alternatives are not correlated.

The most famous example of the IIA violation is the red bus/blue bus paradox. Consider a situation where you are choosing between traveling by car, train, or blue bus. For the sake of simplicity, we assume that the probability of selecting each of the options is equal to 1/3. Under IIA, if we added a red bus to the choice set, the ratio of the probabilities of the remaining options would remain constant so the probabilities would now equal 1/4.

However, in reality, does the color of the bus matter that much?! Let's, for the sake of argument, assume that it does not, and in effect we are still selecting between the car, train, or bus. Therefore, adding a new bus option should not alter the real probabilities of selecting a car or train and these should still be 1/3. However, the probabilities of selecting either blue or red buses would split equally, so each of the bus options would have 1/6 chances of being selected.

This violates the IIA property: while the ratio *P(car) / P(train)* did not change, the ratio between *P(car) / P(blue bus)* as well as *P(train) / P(blue bus)* did.

In this recipe, we will see if the IIA in our dataset is violated. In the next two recipes, we will present models that are capable of handling the correlation between alternatives.

Getting ready

To execute this recipe, you need a working Python Biogeme package installed on your machine. No other prerequisites are required.

How to do it...

Having estimated the MNL model in the previous recipe, let's calculate the probabilities of choosing the alternatives given the estimates. We will then re-estimate the model with the first alternative removed and repeat the procedure. Finally, we will compare the ratios of probabilities that we get from each model. Files for this recipe are located in the `TestingIIA` folder.

The simulation model specification script is almost identical to the estimation one so we will only discuss the differences here (the `TestingIIA/dcm_mnl_simul.py` file):

```
C_price = Beta('C_price',-7.29885,-10,10,0,'C price' )
V_price = Beta('V_price',-5.07495,-10,10,0,'V price' )
Y_price = Beta('Y_price',-4.40754,-10,10,0,'Y price' )
Z_price = Beta('Z_price',-8.70638,-10,10,0,'Z price' )

ASC = Beta('ASC',0,-10,10,1,'ASC' )
B_comp = Beta('B_comp',3.52571,-10,10,0,'compartment' )
B_refund = Beta('B_refund',-0.718748,-3,3,0,'refund' )


. . .


# availability flags
availability = {
    1:   AA777_1_C_AV,
    . . .
    16: UA110_4_V_AV
}

# The choice model is a logit, with availability conditions
probAA777_C = bioLogit(V, availability, 1)
. . .
probUA110_V = bioLogit(V, availability, 16)

# Defines an iterator on the data
rowIterator('obsIter')

# exclude observations where AA777 C was selected
exclude = choice == 1
BIOGEME_OBJECT.EXCLUDE = exclude

# simulate
simulate = {
    'P_AA777_C': probAA777_C,
    . . .
    'P_UA110_V': probUA110_V
```

```
}

## Code for the sensitivity analysis
names = ['B_comp','B_refund','C_price','V_price','Y_price','Z_price']

values = [[1.71083,-0.0398667,-1.67587,0.190499,0.209566,-2.13821],[-
0.0398667,0.0188657,-0.00717013,-0.083915,-0.0941582,0.0155518],[-
1.67587,-0.00717013,1.76813,0.0330621,0.0365816,2.18927],[0.190499,-
0.083915,0.0330621,0.418485,0.452985,-0.0676863],[0.209566,-
0.0941582,0.0365816,0.452985,0.498726,-0.0766095],[-
2.13821,0.0155518,2.18927,-0.0676863,-0.0766095,2.74714]]

vc = bioMatrix(6,names,values)
BIOGEME_OBJECT.VARCOVAR = vc
BIOGEME_OBJECT.SIMULATE = Enumerate(simulate,'obsIter')

# Statistics
nullLoglikelihood(availability,'obsIter')
...

# Parameters
BIOGEME_OBJECT.PARAMETERS['RandomDistribution'] ='MLHS'
BIOGEME_OBJECT.PARAMETERS['NbrOfDraws'] = '1'


...
```

How it works...

As with the estimation model, we first load all the necessary packages that Python Biogeme uses.

Next, we define the coefficients. However, we use the values of the coefficients that we already estimated. These can be sourced from the dcm_mnl_param.py file in the MNL folder. This is followed by specifying the dictionaries of attributes and utilities.

As we will be excluding the AA777_C alternative in our IIA tests, we can remove all the observations where that alternative was chosen from our full model. All you need to do is specify the condition—in our case, choice == 1—and then set the .EXCLUDE variable in BIOGEME_OBJECT.

 More sophisticated exclusion conditions are, of course, possible.

Once the availability and exclusions are determined, we create a variable for each probability using the `biologit(...)` method and put it in the simulate dictionary. The method takes the dictionary of all utilities as its first argument and dictionary of availability conditions as the second. The last parameter specifies the alternative number.

Lastly, we specify the names of the coefficients and covariance matrix between the coefficients. These can again be sourced from the `dcm_mnl_param.py` file in the `MNL` folder. Python Biogeme uses them in the sensitivity analysis.

Finally, we instruct the package to `.SIMULATE` the probabilities using the `Enumerate(...)` method. The method takes the simulate dictionary as its first parameter and the rows iterator as the second one. As we are interested only in the estimation of probabilities (and not really a Monte Carlo analysis), we specify the `RandomDistribution` to **Modified Latin Hypercube Sampling** (**MLHS**), with only one draw.

We run the simulation just like with the estimation (type the command in one line):

```
pythonbiogeme dcm_mnl_simul ../../../Data/Chapter10/choices.dat
```

The package creates `dcm_mnl_simul.html` that contains a table with the probabilities calculated, as shown in the following screenshot:

Simulation report

Number of draws for Monte-Carlo: 1

Type of draws: MLHS

Number of draws for sensitivity analysis: 100

Row	P AA777_C	P AA777_C_5	P AA777_C_95	P AA777_C_median	P AA777_V	P AA777_V_5	P AA777_V_95	P AA777_V_median	P AA777_Y
1	0.0100628	0.00893329	0.0110846	0.00991616	0.140337	0.134924	0.146008	0.140544	0
2	0.0122077	0.0107808	0.0134883	0.0120666	0.17025	0.164788	0.176655	0.170475	0.0555726
3	0.0123868	0.0109823	0.013687	0.012241	0.172748	0.16618	0.179116	0.172977	0.0563879
4	0.00914733	0.00811727	0.0100794	0.00903639	0.127569	0.122885	0.132641	0.127863	0.0416408
5	0.00884196	0.00784132	0.00974378	0.00873988	0.123311	0.118949	0.128188	0.123547	0.0402507
6	0.0126407	0.0113248	0.0139939	0.0124391	0.176288	0.170665	0.183116	0.176861	0.0575434
7	0.00951026	0.00843063	0.0104772	0.00938694	0.132631	0.127791	0.138122	0.132916	0
8	0.0116849	0.0103362	0.012884	0.0115675	0.162959	0.157118	0.169253	0.16343	0

The table has 10,000 rows. For the sensitivity analysis, the simulation procedure draws 100 uniformly distributed values of the coefficient parameter and returns the 5th, 95th, and median values of the probability for each alternative.

As the last step, we estimate the model with the first alternative removed and repeat the simulation procedure (see the `dcm_iia.py` and `dcm_iia_simul.py` files in the `TestingIIA` folder):

Simulation report

Number of draws for Monte-Carlo: 1

Type of draws: MLHS

Number of draws for sensitivity analysis: 100

Row	P(AA777_V)	P(AA777_V)_5	P(AA777_V)_95	P(AA777_V)_median	P(AA777_Y)	P(AA777_Y)_5	P(AA777_Y)_95	P(AA777_Y)_medi
1	0.141826	0.137788	0.145165	0.141558	0	0	0	0
2	0.172479	0.166469	0.177376	0.171985	0.0562227	0.054192	0.0576525	0.0562822
3	0.174981	0.169617	0.179689	0.174642	0.0570383	0.0552599	0.0583564	0.0569088
4	0.128826	0.125004	0.131913	0.128539	0.0419933	0.0406623	0.0430568	0.0419313
5	0.124495	0.120694	0.127556	0.124196	0.0405814	0.0392557	0.0415943	0.0405322
6	0.178566	0.174832	0.181368	0.178322	0.0582066	0.0564886	0.059829	0.0581229
7	0.13402	0.129855	0.137288	0.133686	0	0	0	0
8	0.165013	0.159553	0.169521	0.164562	0	0	0	0
9	0.139177	0.134864	0.142664	0.138787	0.0453671	0.0437886	0.0465908	0.0453416
10	0.124495	0.120694	0.127556	0.124196	0.0405814	0.0392557	0.0415943	0.0405322

As you can see, the `AA777_C` alternative is no longer available. Now, if the IIA postulate is not violated, we will see no change in the ratios of probabilities. Let's analyze the choice situation from the second row.

`P(AA777_V)` from the original model was `0.17025`, and `P(AA777_Y)` was `0.0555726`; thus, their ratio was 3.0636. In the new model, we have `P(AA777_V) = 0.172479` and `P(AA777_Y) = 0.0562227`; the new ratio is therefore 3.0678.

Not a big change.

There's more...

This is, however, only one observation. Let's check it for the rest of the choice situations (the `dcm_iia_testing.py` file):

```
import pandas as pd

# read the html tables
old_model = 'dcm_mnl_simul.html'
new_model = 'dcm_iia_simul.html'

old_model_p = pd.read_html(old_model, header = 0)[3]
new_model_p = pd.read_html(new_model, header = 0)[3]

# let's look at only two columns
```

```
cols = ['P_AA777_Y', 'P_AA777_V']

# make sure that there are no zeros
old_model_p = old_model_p[cols]
old_model_p = old_model_p[old_model_p[cols[0]] != 0]
old_model_p = old_model_p[old_model_p[cols[1]] != 0]

new_model_p = new_model_p[cols]
new_model_p = new_model_p[new_model_p[cols[0]] != 0]
new_model_p = new_model_p[new_model_p[cols[1]] != 0]

# calculate the ratios
old_model_p['ratios_old'] = old_model_p \
    .apply(lambda row: row['P_AA777_V'] / row['P_AA777_Y'],
        axis=1)

new_model_p['ratios_new'] = new_model_p \
    .apply(lambda row: row['P_AA777_V'] / row['P_AA777_Y'],
        axis=1)

# join with the old model results
differences = old_model_p.join(new_model_p, rsuffix='_new')

# and calculate the differences
differences['diff'] = differences\
    .apply(lambda row: row['ratios_new'] - row['ratios_old'],
        axis=1)

# calculate the descriptive stats for the columns
print(differences[['ratios_old', 'ratios_new', 'diff']] \
    .describe())
```

We use `pandas` as it provides out-of-the-box methods to read in the tables from HTML files (refer to the *Retrieving HTML pages with Pandas* recipe in *Chapter 1, Preparing the Data*).

Once we read the files in, we select only the two columns that we tested earlier from both the full model and reduced one, and make sure that there are no zeros in either of the columns. Next, we calculate the ratios and join the two files so that we can calculate the differences between corresponding observations. Finally, we produce descriptive statistics for the ratios and the ratios difference columns:

	ratios_old	ratios_new	diff
count	6652.000000	6652.000000	6652.000000
mean	3.063567	3.067787	0.004220
std	0.000006	0.000006	0.000008

min	3.063548	3.067773	0.004196
25%	3.063562	3.067783	0.004211
50%	3.063568	3.067785	0.004219
75%	3.063574	3.067791	0.004225
max	3.063580	3.067803	0.004247

The differences do not change that much—the coefficient of variation (standard deviation / average) is close to 0 and the maximum difference is 0.004247. Therefore, we cannot conclude that the IIA is violated.

Handling IIA violations with the Nested Logit model

If, however, your models violate the IIA, you need to resort to more advanced models. In this recipe, we will see one of them—a slightly more complicated Nested Logit model. The model groups similar alternatives into nests (hence the name). Given the limited space here, we will not discuss the intricacies of the model, but I highly recommend Kenneth Train's book that I referred to earlier in this chapter.

Getting ready

To execute this recipe, you need a working Python Biogeme package installed on your machine. No other prerequisites are required.

How to do it...

The skeleton of the model code remains the same; here, we will only show the changes (the Nested/dcm_nested.py file):

```
# add the coefficients to be estimated
C_price = Beta('C_price',0,-10,10,0,'C price' )
V_price = Beta('V_price',0,-10,10,0,'V price' )
Y_price = Beta('Y_price',0,-10,10,0,'Y price' )
Z_price = Beta('Z_price',0,-10,10,0,'Z price' )

ASC = Beta('ASC',0,-10,10,1,'ASC' )

B_refund = Beta('B_refund',0,-3,3,0,'refund' )
B_comp = Beta('B_comp',0,-3,3,0,'compartment' )

# nest mu parameter
biz_mu = Beta('biz_mu',0.5,0,1,0)
```

```
eco_mu = Beta('eco_mu',0.5,0,1,0)


...


# 1: nests parameter
# 2: list of alternatives
business = biz_mu, [1,2,5,6,9,10,13,14]
economy  = eco_mu, [3,4,7,8,11,12,15,16]
nests = business, economy

# The choice model is a logit, with availability conditions
logprob = lognested(V, availability, nests, choice)

# Defines an itertor on the data
rowIterator('obsIter')

# DEfine the likelihood function for the estimation
BIOGEME_OBJECT.ESTIMATE = Sum(logprob,'obsIter')

# Statistics

nullLoglikelihood(availability,'obsIter')
choiceSet = [1,2,3,4,5,6,7,8,9,10,11,12,13,14,15,16]
cteLoglikelihood(choiceSet,choice,'obsIter')
availabilityStatistics(availability,'obsIter')

BIOGEME_OBJECT.PARAMETERS['optimizationAlgorithm'] = 'BIO'
BIOGEME_OBJECT.PARAMETERS['checkDerivatives'] = '1'
BIOGEME_OBJECT.PARAMETERS['numberOfThreads'] = '8'
BIOGEME_OBJECT.PARAMETERS['moreRobustToNumericalIssues'] = '0'
...
```

How it works...

The main difference between MNL and NL is, as already pointed out, the fact that NL groups alternatives that share common characteristics into nests. Each nest is associated with a lambda parameter: a parameter that defines how strong the competition is between alternatives within the nest. The lambda parameter takes values between 0 (highest competition) and 1 (no competition).

In our example, we assume that the business alternatives (the C and Z booking classes) both would enjoy larger legroom and better service so we group them in the business nest; the two economy alternatives (the Y and V booking classes) are grouped in the economy nest. The biz_mu and eco_mu are our lambda parameters for the nests.

Next, we determine which alternatives belong to which nest by passing a list with alternative numbers and the `lambda` parameter.

We also use a different estimator—the `lognested(...)` method, in contrast to `bioLogLogit(...)`; this additionally includes the `nests` object:

Estimation report

Number of estimated parameters: 8
Sample size: 10000
Excluded observations: 0
Init log likelihood: –25709.877
Final log likelihood: –21617.456
Likelihood ratio test for the init. model: 8184.842
Rho-square for the init. model: 0.159
Rho-square-bar for the init. model: 0.159
Final gradient norm: +4.447e+01
Diagnostic: Convergence reached...
Iterations: 18
Run time: 00:51
Nbr of threads: 8

Estimated parameters

Click on the headers of the columns to sort the table [Credits]

Name	Value	Std err	t-test	p-value		Robust Std err	Robust t-test	p-value	
B_comp	-0.673	0.441	-1.53	0.13	*	0.451	-1.49	0.14	*
B_refund	-0.617	0.131	-4.71	0.00		0.131	-4.69	0.00	
C_price	-3.13	0.698	-4.49	0.00		0.705	-4.45	0.00	
V_price	-5.53	0.623	-8.88	0.00		0.625	-8.85	0.00	
Y_price	-4.92	0.678	-7.25	0.00		0.679	-7.24	0.00	
Z_price	-3.45	0.719	-4.80	0.00		0.729	-4.73	0.00	
biz_mu	1.00	1.80e+308	0.00	1.00	*	1.80e+308	0.00	1.00	*
eco_mu	1.00	1.80e+308	0.00	1.00	*	1.80e+308	0.00	1.00	*

As expected, the nesting structure brought nothing useful to the model. The Rho-squared got slightly better but at the expense of two additional degrees of freedom. In addition, `biz_mu` and `biz_eco` are not statistically significant.

Managing sophisticated substitution patterns with the Mixed Logit model

The Mixed Logit model, in contrast to all the previously presented models, allows some of the coefficients to be random following a normal distribution, that is, having a mean and standard deviation. This, in effect, eliminates the dependency on the IIA assumption and allows the flexible modeling of substitution patterns. However, this comes at the cost of computation time.

Getting ready

To execute this recipe, you need a working Python Biogeme package installed on your machine. No other prerequisites are required.

How to do it...

As we have already established that the MNL model estimated using our dataset does not violate the IIA property, we will only present the mechanics of estimating the Mixed Logit model (the `MixedLogit/dcm_mixed.py` file):

```
C_price = Beta('C_price',0,-10,10,0,'C price' )
V_price = Beta('V_price',0,-10,10,0,'V price' )
Y_price = Beta('Y_price',0,-10,10,0,'Y price' )
Z_price = Beta('Z_price',0,-10,10,0,'Z price' )

ASC = Beta('ASC',0,-10,10,1,'ASC' )

B_ref = Beta('B_ref',0,-3,3,0,'refund' )
B_ref_S = Beta('B_ref_S',0,-3,3,0,'refund (std)' )
B_comp = Beta('B_comp',0,-3,3,0,'compartment' )

# Random parameters
B_ref_rnd = B_ref + B_ref_S * bioDraws('B_ref_rnd')
...

# The dictionary of utilities is constructed.
V = {}

V[1] = Z_price * p[1] + B_ref_rnd * r[1] + B_comp * c[1]
V[16] = Y_price * p[16] + B_ref_rnd * r[16] + B_comp * c[16]

# availability flags
availability = {
```

```
      1:   AA777_1_C_AV,
      ...
      16:  UA110_4_V_AV
  }


  # The choice model is a logit, with availability conditions
  prob = bioLogit(V, availability, choice)
  l = mixedloglikelihood(prob)


  # Defines an itertor on the data
  rowIterator('obsIter')


  # Define the likelihood function for the estimation
  BIOGEME_OBJECT.ESTIMATE = Sum(l, 'obsIter')


  # Statistics

  nullLoglikelihood(availability,'obsIter')
  choiceSet = [1,2,3,4,5,6,7,8,9,10,11,12,13,14,15,16]
  cteLoglikelihood(choiceSet, choice, 'obsIter')
  availabilityStatistics(availability, 'obsIter')


  BIOGEME_OBJECT.PARAMETERS['NbrOfDraws'] = '100'
  BIOGEME_OBJECT.DRAWS = { 'B_ref_rnd': 'NORMAL' }
  ...
```

How it works...

The random `B_ref_rnd` coefficient consists of the deterministic part (average) `B_ref` and the stochastic (random) part `B_ref_S`—the standard deviation.

The `bioDraws(...)` method is used to produce the random draws. The draws are generated for each observation and that is where the performance hit comes from: for each observation, the estimator draws a predefined number of samples from a normal distribution in order to estimate the coefficient.

The `random` parameter is then included in each utility function instead of `B_ref`; doing so allows the handling of the substitution patterns, as the model will handle the correlation between the alternatives along this dimension.

The `mixedloglikelihood(...)` method handles the estimation of the model. It requires a `bioLogit(...)` object as its only parameter.

Finally, we set the .DRAWS variable to define what distribution to draw from; in our example, we draw from a normal distribution. The results of the model are presented below:

Estimation report

Number of draws: 100
Number of estimated parameters: 7
Sample size: 10000
Excluded observations: 0
Init log likelihood: -25531.498
Final log likelihood: -21617.446
Likelihood ratio test for the init. model: 7828.105
Rho-square for the init. model: 0.153
Rho-square-bar for the init. model: 0.153
Final gradient norm: +7.458e-04
Diagnostic: Convergence reached...
Iterations: 6
Run time: 03:07
Nbr of threads: 8

Estimated parameters

Click on the headers of the columns to sort the table [Credits]

Name	Value	Std err	t-test	p-value		Robust Std err	Robust t-test	p-value	
B_comp	-0.673	0.441	-1.53	0.13	*	0.451	-1.49	0.14	*
B_ref	-0.618	0.131	-4.71	0.00		0.131	-4.70	0.00	
B_ref_S	0.0497	0.340	0.15	0.88	*	0.0951	0.52	0.60	*
C_price	-3.13	0.698	-4.49	0.00		0.705	-4.45	0.00	
V_price	-5.53	0.623	-8.88	0.00		0.625	-8.85	0.00	
Y_price	-4.92	0.678	-7.25	0.00		0.679	-7.24	0.00	
Z_price	-3.45	0.719	-4.80	0.00		0.729	-4.73	0.00	

As expected, the additional estimated coefficients do not add value to our model; B_ref_S is not statistically significant and can be removed from the model without losing accuracy.

Compared to our previous models, you can see that ML takes the longest to estimate: the MNL takes 1 second to estimate whereas the NL takes 50 seconds in comparison.

11

Simulations

In this chapter, you will learn the following recipes:

 ▶ Using SimPy to simulate the refueling process of a gas station
 ▶ Simulating out-of-energy occurrences for an electric car
 ▶ Determining if a population of sheep is in danger of extinction due to a wolf pack

Introduction

Obtaining data is always cumbersome: collected data is almost always dirty and requires lots of work to extract the features that you are after. Also, collected data is almost always myopic in its scope: you observe only a portion of all the interactions that happen in any given environment.

However, you can simulate certain situations. Simulations come in handy when, among other things, it is impossible to observe every single part of the environment, if you want to test your models in various situations, or you want to validate your assumptions.

A number of other books will teach you simulations of financial data. In this book, we will not be doing this. In contrast, we will focus on agent-based simulations. This type of simulation creates a virtual world (or environment) where we place our agents. Agents can represent almost anything that you can think of: in our simulations, an agent will be a gas station, car, recharge station, or a sheep and wolf. Throughout the simulation, agents interact with other agents and environments that alter their behavior.

However, very powerful, agent-based simulations have some limitations. The main one comes from the limitations (or accuracy) of interactions between agents; missing out on an important behavior will almost inevitably lead to wrong conclusions. For example, a simple assumption that drivers on the road always behave logically might lead to a conclusion that a roundabout needs only one line, whereas in reality, simply because people's behavior varies, it would cause a mile-long traffic jam. If you want to read more about agent-based simulations and their limitations, I can recommend this article at `http://ijcsi.org/papers/IJCSI-9-1-3-115-119.pdf`.

This chapter will teach you how to design and run simulations using `SimPy`.

I strongly recommend reading the introduction to modeling with `SimPy` on their website:

`https://simpy.readthedocs.org/en/latest/topical_guides/index.html`

You can download the package from `http://pypi.python.org/pypi/SimPy/`. Once downloaded, you can install the package by issuing the following statement from the command line:

```
gunzip simpy-3.0.8.tar.gz
tar xvfz simpy-3.0.8.tar
cd simpy-3.0.8
python setup.py install
```

If you are running Anaconda, you can also use the following:

```
pip install -U simpy
```

In this chapter, we will also touch on the more sophisticated parts of Python: writing and inheriting classes and interrupting the execution of the simulation with exceptions. Don't worry, we will walk through the examples presented in detail.

Using SimPy to simulate the refueling process of a gas station

If you own a car, visiting a gas station every other week has most likely become part of your routine: pull over next to the pump, swipe your card (or pay inside), fuel up your car, and leave.

However, if you put yourself in the shoes of a gas station owner before it is even built, a number of possible questions come to mind:

- What is the frequency of cars pulling over to fill up?
- How would that affect the wait time for each driver? (We do not want a driver to wait for too long as he or she might choose a competitor.)
- Will the number of distributors be enough to satisfy demand?
- How often will I have to call the supplier to replenish the tanks?
- How much profit can I make?

These are, however, just a handful of questions that you, as the owner of a future gas station, might be asking yourself.

There are also those *what-if* questions. *What if I add two more distributors? What if I put bigger tanks so that I do not have to call the supplier so often? What if the frequency of cars arriving at my gas station is not every 30 seconds but every 45 seconds? How will that affect my bottom line?*

Simulations can be a really inexpensive and invaluable tool in testing a variety of different scenarios and situations before you invest millions of your dollars.

Note that the flow of the chapter differs to what we have been presenting in the earlier parts of the book.

Getting ready

To execute this recipe, you will need `SimPy` and `NumPy`. No other prerequisites are required.

How to do it...

The simulation code is lengthy but having gone through the whole book it should be fairly straightforward by now. It might be, however, a little bit awkward so use the source code from GitHub as a reference if something is not clear—it is located in the `sim_gasStation.py` file. We will go through it step by step in the next section:

```python
import numpy as np
import simpy
import itertools


if __name__ == '__main__':
    # what is the simulation horizon (in seconds)
    SIM_TIME = 10 * 60 * 60 # 10 hours

    # create the environment
```

```
env = simpy.Environment()

# create the gas station
gasStation = GasStation(env)

# create the process of generating cars
env.process(Car.generate(env, gasStation))

# and run the simulation
env.run(until = SIM_TIME)
```

How it works...

As always, we start by importing all the necessary modules first.

The execution of our script starts at the first line after `if __name__ == '__main__'` (by Python's convention). First, we determine how long our simulation should run for. For the sake of this example, we treat each epoch of the simulation to be 1 second. Next, as will be the case with every simulation that we run, we create `.Environment(...)`. This is the basis of our simulation. The environment encapsulates time and handles the interactions between all the processes and agents in our simulation.

Next, we create our first agent, `GasStation`:

```
class GasStation(object):
    def __init__(self, env):
        # keep a pointer to the environment in the class
        self.env = env

        # fuel capacity (gallons) and reservoirs
        # to track level
        self.CAPACITY = {'PETROL': 8000, 'DIESEL': 3000}
        self.RESERVOIRS = {}
        self.generateReservoirs(self.env, self.CAPACITY)

        # number of pumps for each fuel type
        self.PUMPS_COUNT = {'PETROL': 3, 'DIESEL': 1}
        self.PUMPS = {}
        self.generatePumps(self.env, self.CAPACITY,
            self.PUMPS_COUNT)

        # how quickly they pump the fuel
        self.SPEED_REFUEL = 0.3 # 0.3 gal/s

        # set the minimum amount of fuel left before
        # replenishing
```

```
self.MINIMUM_FUEL = {'PETROL': 300, 'DIESEL': 100}

# how long does it take for the truck to get
# to the station after placing the call
self.TRUCK_TIME = 200
self.SPEED_REPLENISH = 5

# add the process to control the levels of fuel
# available for sale
self.control = self.env.process(self.controlLevels())

print('Gas station generated...')
```

Every class in Python that you want needs to have an __init__(self, ...) method. This method is called when you want to create a GasStation object (in this example).

> You can have a class with only static methods and then you do not need __init__(self, ...). A static method is a method that does not require an object of the class to execute. In other words, think about a method that you can call that does not depend on any property of a class object but is still tied to the theme of the class.
>
> For example, you might have a Triangle class and create a length method to calculate the length of the hypotenuse using the Pythagorean theorem. However, you can use the same method to calculate the distance between two points on a plane (Cartesian coordinates). Thus, you might want to make the method static and reuse it in other parts of your code when you need to calculate the length of a line between two points.

You have probably noticed by now that the word self appears in (almost) every other method. The self parameter is a reference to the instantiated object itself (hence the name).

Within the __init__(self, ...) method, you should list all the internal attributes that each of the objects instantiated with such a class will have. Our gas station will offer two types of fuel: petrol and diesel. The underground tanks will hold, respectively, 8,000 and 3,000 gallons of fuel each. As our gas station will offer two types of fuel, the self.RESERVOIRS attribute will hold two .Containers(...): one for each fuel type. The .Container(...) object is effectively a resource that can be shared by many processes; each process can access this common supply and use it until either the process finishes or there is nothing left in .Container(...). We generate these using the .generateReservoirs(...) method:

```
def generateReservoirs(self, env, levels):
    for fuel in levels:
        self.RESERVOIRS[fuel] = simpy.Container(
            env, levels[fuel], init=levels[fuel])
```

The `.Container(...)` method takes the pointer to the `.Environment(...)` object as its first argument, the size of the reservoir as the second one, and the initial level as the third.

Next, we create `FuelPump` objects using the `generatePumps(...)` method:

```
def generatePumps(self, env, fuelTypes, noOfPumps):
    '''
        Helper method to generate pumps
    '''
    for fuelType in fuelTypes:
        self.PUMPS[fuelType] = FuelPump(
            env, noOfPumps[fuelType])
```

The only `FuelPump` object attribute is `self.resource` that will hold the `.Resource(...)` object of `SimPy`. The `.Resource(...)` object can be thought of as an ultimate gatekeeper of certain resources. It can be used to limit the number of parallel processes that access a certain `.Container(...)`.

The easiest (in my opinion) way to understand the distinction between `.Container(...)` and `.Resource(...)` is to think about how any gas station works. There are multiple distributors (with fuel nozzles) connected to the same reservoir (big underground tank) with fuel. At any given time, a distributor can be used by only one car; thus, the throughput of a gas station is limited by the number of distributors that the gas station has. That is why we treat `FuelPump` as `.Resource(...)`. Also, as all the distributors are connected to the same reservoir, we use the `.Container(...)` object to model that behavior. Each time a car arrives at the gas station, we will use the `.request(...)` method of `.Resource(...)` to get access to one of the distributors; if all the distributors are currently being used, we will have to wait until one becomes available.

Next, we specify the speed of refueling—in our case, it is 0.3 gal/second—and the minimum levels of the reservoirs before we call the truck with supplies; for petrol, we call the truck when the level gets at or below 300 gallons, and for diesel, when we reach 100 gallons or less. Also, what we need to include in our calculations is how long it takes the truck to get to the station (we assume 300 seconds in this simplified example) and the speed of replenishing the fuel at 5 gallons per second.

Finally, the last command within the `__init__(self, ...)` method creates the first process in the environment; we put the `.controlLevels()` method in the environment:

```
def controlLevels(self):
    while True:
        # loop through all the reservoirs
        for fuelType in self.RESERVOIRS:

            # and if the level is below the minimum
            if self.RESERVOIRS[fuelType].level \
```

```
< self.MINIMUM_FUEL[fuelType]:

    # replenishes
    yield env.process(
        self.replenish(fuelType))
# wait 5s before checking again
yield env.timeout(5)
```

The method runs an infinite loop and will run for the whole duration of the simulation. Within the loop, it checks the levels of all the reservoirs every 5 seconds; we use the yield `env.timeout(5)` call to the environment to check only every 5 seconds.

Throughout the whole simulation span, various agents will create processes and then yield them. The yield command suspends the process and gets triggered only by the environment or another process. In our example, yield `env.timout(5)` suspends the current process for 5 seconds; after 5 seconds have elapsed, the environment will trigger another iteration through the loop.

In the loop, we go thorough our reservoirs and check whether the amount of fuel left fell below the critical level. If so, we yield this process to call the truck, that is, the `replenish(...)` method. The current loop will only resume (be triggered back) after the replenishing is finalized:

```
def replenish(self, fuelType):
    # waiting for the truck to come (lead time)
    yield self.env.timeout(self.TRUCK_TIME)

    # how much we need to replenish
    toReplenish = self.RESERVOIRS[fuelType].capacity - \
        self.RESERVOIRS[fuelType].level

    # wait for the truck to dump the fuel into
    # the reservoirs
    yield (self.env.timeout(toReplenish
        / self.SPEED_REPLENISH))

    # and then add the fuel to the available one
    yield self.RESERVOIRS[fuelType].put(toReplenish)
```

 Note that, in some circumstances, the multiline commands separated by \ might produce an error. You can avoid this by enclosing the multiline commands within parentheses (...).

In the `replenish(...)` method, we first yield the process to wait for the truck to arrive. Once the process resumes, we check how much fuel needs to be replenished. The `.Container(...)` object has two attributes that we call: `.capacity` and `.level`. The `.capacity` attribute returns the capacity of the reservoir and `.level` holds the current amount of fuel in the reservoir. The difference between the two shows how much needs to be replenished. Knowing the amount of fuel required and the speed of replenishing, we calculate the time that the process will take to finish and yield this process for that time. Once the wait time has elapsed, we `.put(...)` the required amount of fuel in `.Container(...)` and return the execution to the caller method.

Finally, we specify the `Car` class:

```
class Car(object):
    def __init__(self, i, env, gasStation):
        # pointers to the environment and gasStation objects
        self.env = env
        self.gasStation = gasStation

        # fuel type required by the car
        self.FUEL_TYPE = np.random.choice(
            ['PETROL', 'DIESEL'], p=[0.7, 0.3])

        # details about the car
        self.TANK_CAPACITY = np.random.randint(12, 23) # gal

        # how much fuel left
        self.FUEL_LEFT = self.TANK_CAPACITY \
            * np.random.randint(10, 40) / 100

        # car id
        self.CAR_ID = i

        # start the refueling process
        self.action = env.process(self.refuel())
```

Our cars run on specific types of fuel and have a predetermined tank capacity. When we create a `car` object, it also has a randomly assigned amount of fuel left in the tank—between 10% and 40% of the nominal tank capacity,

The process starts with refueling:

```
def refuel(self):
    # what's the fuel type so we request the right pump
    fuelType = self.FUEL_TYPE

    # let's get the pumps object
```

```
pump = gasStation.getPump(fuelType)

# and request a free pump
with pump.request() as req:
    # time of arrival at the gas station
    arrive = self.env.now

    # wait for the pump
    yield req

    # how much fuel does the car need
    required = self.TANK_CAPACITY - self.FUEL_LEFT

    # time of starting refueling
    start = self.env.now
    yield self.gasStation.getReservoir(fuelType)\
        .get(required)

    # record the fuel levels
    petrolLevel = self.gasStation\
                .getReservoir('PETROL').level
    dieselLevel = self.gasStation\
                .getReservoir('DIESEL').level

    # and wait for it to finish
    yield env.timeout(required / gasStation \
        .getRefuelSpeed())
```

First, we retrieve all the pumps for the specific fuel type from the `GasStation` object. The fuel type is our `.Resource(...)` that we created earlier: it is a single object for each fuel type with a specified number of available pumps—three for petrol and only one for diesel. Each time a car arrives at the station, it uses the `.request(...)` object to access a pump. We then wait for the request to be fulfilled.

Once a pump becomes available, the execution resumes. We first calculate how much fuel is required to fill up the tank and begin the refueling process. The process yields for the amount of time it takes to fill up the required amount of gas.

This, in its entirety, is the whole refueling process that each car and the gas station will go through.

The `generate(...)` static method generates cars at random (one car every 5 to 45 seconds) throughout the simulation timespan:

```
@staticmethod
def generate(env, gasStation):
    '''
```

```
        A static method to generate cars
'''
# generate as many cars as possible during the
# simulation run
for i in itertools.count():
    # simulate that a new car arrives between 5s
    # and 45s
    yield env.timeout(np.random.randint(5, 45))

    # create a new car
    Car(i, env, gasStation)
```

We add the car generation process to the environment using the `env.process(Car.generate(env, gasStation))` command.

 As you can see, the static method, in contrast to the object method, does not have the `self` keyword.

Finally, we call the `.run(...)` method. The `until` parameter specifies how long to run the simulation for.

Once you execute the script, you will see something as follows:

```
Gas station generated...

                                                   Left
CarID   Arrive  Start   Finish  Gal     Type      Petrol  Diesel
----------------------------------------------------------------
0       6       6       54      14.60   PETROL    7985    3000
1       27      27      57      9.24    PETROL    7976    3000
2       42      42      89      14.28   DIESEL    7976    2985
3       75      75      127     15.75   PETROL    7960    2985
4       87      87      152     19.58   PETROL    7940    2985
5       129     129     168     11.70   PETROL    7929    2985
6       141     141     197     16.80   PETROL    7912    2985
7       178     178     209     9.48    DIESEL    7912    2976
8       205     205     258     16.06   PETROL    7896    2976
9       233     233     279     14.08   DIESEL    7896    2962
10      273     273     314     12.54   PETROL    7883    2962
11      304     304     358     16.34   DIESEL    7883    2945
12      334     334     391     17.20   PETROL    7866    2945
```

Throughout the simulation you will `notice (...)` when the truck is ordered to replenish the fuel and when the whole process finishes:

```
791    20413   20413   20449   11.04   PETROL   784   115
792    20449   20449   20481   9.76    DIESEL   784   105
793    20486   20486   20518   9.80    PETROL   774   105
----------------------------------------------------------------
CALLING TRUCK AT 20540s.
----------------------------------------------------------------
795    20531   20531   20562   9.38    DIESEL   758   96
794    20516   20516   20571   16.60   PETROL   758   105
796    20563   20563   20597   10.37   PETROL   747   96
797    20600   20600   20644   13.32   PETROL   734   96
798    20643   20643   20677   10.40   PETROL   723   96
799    20686   20686   20724   11.48   PETROL   712   96
800    20703   20703   20732   8.88    PETROL   703   96
----------------------------------------------------------------
TRUCK ARRIVING AT 20740s
TO REPLENISH 2912 GALLONS OF DIESEL
----------------------------------------------------------------
801    20727   20727   20755   8.54    DIESEL   703   87
802    20760   20760   20815   16.72   DIESEL   703   70
803    20776   20776   20816   12.06   PETROL   691   70
804    20812   20812   20843   9.48    PETROL   682   70
805    20822   20822   20864   12.64   PETROL   669   70
806    20830   20830   20880   15.00   PETROL   654   70
807    20850   20850   20896   13.86   PETROL   640   70
808    20864   20864   20902   11.55   PETROL   629   70
810    20892   20896   20921   7.68    PETROL   604   70
809    20875   20880   20937   17.22   PETROL   611   70
811    20926   20926   20972   14.00   PETROL   590   70
812    20951   20951   20982   9.36    PETROL   580   70
813    20960   20960   20991   9.49    PETROL   571   70
814    20998   20998   21028   9.10    PETROL   562   70
815    21024   21024   21062   11.48   DIESEL   562   59
816    21057   21057   21096   11.88   PETROL   550   59
817    21062   21062   21104   12.75   PETROL   537   59
818    21102   21102   21135   10.08   DIESEL   537   49
819    21121   21121   21161   12.18   PETROL   525   49
820    21164   21164   21196   9.62    PETROL   515   49
821    21180   21180   21242   18.69   PETROL   497   49
822    21214   21214   21254   12.00   PETROL   485   49
823    21237   21237   21279   12.80   DIESEL   485   36
----------------------------------------------------------------
FINISHED REPLENISHING AT 21322s.
----------------------------------------------------------------
824    21274   21274   21331   17.20   PETROL   467   36
```

There's more...

What about the profits? In this part of our recipe, we will assume that the gas station buys petrol for $1.95 and sells for $2.45 per gallon, and buys diesel for $1.67, which it sells for $2.23 per gallon. We also added the initial cost of purchasing the fuel to fill up the tanks (the `sim_gasStation_alternative.py` file):

```
# cash registry
self.sellPrice = {'PETROL': 2.45, 'DIESEL': 2.23}
self.buyPrice  = {'PETROL': 1.95, 'DIESEL': 1.67}
self.cashIn = 0
self.cashOut = np.sum(
    [ self.CAPACITY[ft] \
    * self.buyPrice[ft]
    for ft in self.CAPACITY])
self.cashLost = 0
```

In the `replenish(...)` method, we also added a line to account for the fact that we need to pay for the fuel ordered:

```
self.pay(toReplenish * self.buyPrice[fuelType])
```

The `.pay(...)` method increases the amount of money in the `self.cashOut` variable that is paid to the supplier.

We also assume that some of the customers would grow impatient and leave after waiting at the pump for some time (we assumed 5 minutes—see the `waitedTooLong` Boolean variable in the following code). To this end, the Car's `refuel(...)` method has been altered to read as follows:

```
def refuel(self):
    # what's the fuel type so we request the right pump
    fuelType = self.FUEL_TYPE

    # let's get the pumps object
    pump = gasStation.getPump(fuelType)

    # and request a free pump
    with pump.request() as req:
        # time of arrival at the gas station
        arrive = self.env.now

        # wait for the pump
        yield req

        # how much fuel does the car need
```

```
required = self.TANK_CAPACITY - self.FUEL_LEFT

# how long have been waiting for
waitedTooLong = self.env.now - arrive > 5 * 60

if waitedTooLong:
    # leave
    gasStation.lost(required * self.gasStation\
        .getFuelPrice(fuelType))
else:
    # time of starting refueling
    start = self.env.now
    yield self.gasStation.getReservoir(fuelType)\
        .get(required)

    # and wait for it to finish
    yield env.timeout(required / gasStation \
        .getRefuelSpeed())

    # time finished refueling
    fin = self.env.now

    # pay
    toPay = required * self.gasStation\
        .getFuelPrice(fuelType)
    self.gasStation.sell(toPay)

    yield env.timeout(np.random.randint(15, 90))
```

When a car has been queuing for too long, it leaves:

```
113     2747    2986    3043    9.38    DIESEL  6859    2681    $20.92
121     2966    2966    3047    7.68    PETROL  6859    2691    $18.82
122     3010    3010    3110    15.40   PETROL  6844    2681    $37.73
114     2769    3043    3156    10.24   DIESEL  6844    2671    $22.84
----------------------------------------------------------------------
CAR 116 IS LEAVING -- WAIT TOO LONG
----------------------------------------------------------------------
123     3052    3052    3158    10.05   PETROL  6834    2671    $24.62
125     3086    3086    3199    12.60   PETROL  6821    2671    $30.87
129     3163    3163    3260    17.40   PETROL  6788    2656    $42.63
```

You will also notice that we added new methods, `.getFuelPrice(...)` and `.sell(...)`, to the `GasStation` class. These are used to calculate the amount and register the sale of fuel; effectively, the `.sell(...)` method increases the amount of money in the `self.cashIn` variable.

Finally, after the whole simulation finishes, we print out the results to the screen:

```
def printCashRegister(self):
    print('\nTotal cash in:    ${0:8.2f}'\
        .format(self.cashIn))
    print('Total cash out:   ${0:8.2f}'\
        .format(self.cashOut))
    print('Total cash lost: ${0:8.2f}'\
        .format(self.cashLost))
    print('\nProfit: ${0:8.2f}'\
        .format(self.cashIn - self.cashOut))
    print('Profit (if no loss of customers): ${0:8.2f}'\
        .format(self.cashIn - self.cashOut \
            + self.cashLost))
```

The preceding method produces the following output to be printed to the screen:

```
Total cash in:    $82853.85
Total cash out:   $80970.28
Total cash lost: $ 7382.51


Profit: $ 1883.57
Profit (if no loss of customers): $ 9266.08
```

So, as you can see, if the gas station had more fuel distributors, it would have earned over $7,300 more. Based on these simulation parameters, during the first 10 hours, it only made a profit of $154.74. By running the simulation for 20 hours, the profit increased to $1,883.57 but the lost opportunity cost also jumped to $7382.51. This seems like the gas station would require more distributors to satisfy the demand.

Simulating out-of-energy occurrences for an electric car

Electric cars are getting more and more popular these days. However, even though they are cheaper to run, the range of the car somewhat limits its use to travel long distance, at least until a sufficient infrastructure is in place to recharge the car along the way.

In this recipe, we will simulate *out-of-power* situations for an electric car. We start by randomly placing the recharge stations along the way of the car and then simulating the recharge situations. In this recipe, we will allow the driver of the car to drive the car without fully recharging.

Getting ready

To execute this recipe, you will need `SimPy` and `NumPy`. No other prerequisites are required.

How to do it...

As in the previous recipe, we start by defining the environment and all its agents (the `sim_recharge.py` file):

```python
import numpy as np
import simpy


if __name__ == '__main__':
    # what is the simulation horizon (in minutes)
    SIM_TIME = 10 * 60 * 60 # 10 hours

    # create the environment
    env = simpy.Environment()

    # create recharge stations
    rechargeStations = RechargeStation \
        .generateRechargeStations(SIM_TIME)

    # create the driver and the car
    driver = Driver(env)
    car = Car(env, driver, rechargeStations)

    # and run the simulation
    env.run(until = SIM_TIME)
```

How it works...

We start with specifying the parameters of the simulation: we will be driving up to a maximum of 10 hours and will only simulate one car.

First, we create the environment and recharge stations:

```python
@staticmethod
def generateRechargeStations(simTime):
    # we assume an average speed of 35MPH to calculate
    # a maximum distance that might be covered during
    # the simulation timespan
    maxDistance = simTime / 60 * 35 * 2
```

```
# generate the recharge stations
distCovered = 0
rechargeStations = [RechargeStation(env, 0)]

while(distCovered < maxDistance):
    nextStation = np.random.randint(80, 140)

    distCovered += nextStation
    rechargeStations.append(
        RechargeStation(env, distCovered))

return rechargeStations
```

The `.generateRechargeStations(...)` method takes the simulation time horizon and places the recharge stations randomly along the way of the car's travel. The method calculates the maximum distance that a car can travel along the route given the simulation time and assuming an average travel of 35 mph and doubles it. Then, it places one recharge station between 80 and 140 miles.

We add the concept of a driver in this simulation. The driver is responsible for pulling over when he or she passes a recharge station unless it is safe to travel to the next one. He or she also has the authority to break the charging cycle and start driving.

The `drive(...)` method introduces a new concept—an interrupt:

```
def drive(self, car, timeToFullRecharge):
    # decide how long to allow the car to recharge
    interruptTime = np.random.randint(50, 120)

    # if more than the time needed to full recharge
    # wait till the full recharge, otherwise interrupt
    # the recharge process earlier
    yield self.env.timeout(int(np.min(
        [interruptTime, timeToFullRecharge])))

    if interruptTime < timeToFullRecharge:
        car.action.interrupt()
```

As soon as the car pulls over to the recharging station, the driver can stop the recharging and decide to move forward. This is decided at random in the `.drive(...)` method. The method takes the `Car` object as its first parameter and time to fully recharge as the second one. If the randomly generated integer is greater than the time needed for a full recharge, the car will be fully recharged. Otherwise, the recharging will be stopped and the journey will continue.

Interrupting a process in `SimPy` is really straightforward—you just call the `.interrupt()` method. This raises an exception in the process and you can then handle it as an exception.

Most of our simulation code and processing is contained within the `Car` class. At the beginning, we start with defining the `car` parameters. First, we choose between two capacity sizes of batteries, 70 or 85 kWh. Our battery can have a starting level of charge between 80% and 100%. The `.AVG_SPEED` determines how much distance we will be covering during each interval and can range between 36.4 and 49.2 mph. The average economy is stated in kWh per mile; the value of 34 to 38 kWh per 100 miles translates to roughly 100 to 89 mpg-e (miles per gallon equivalent). At the beginning of the trip, we start at location 0.

The simulation begins by starting the process encompassed within the `driving(...)` method:

```
def driving(self):
    # updates every 15 minutes
    interval = 15

    # assuming constant speed -- how far the car travels
    # in each 15 minutes
    distanceTraveled = self.AVG_SPEED / 60 * interval

    # how much battery used to travel that distance
    batteryUsed = distanceTraveled * self.AVG_ECONOMY
```

First, we check how much distance we covered and battery power we used during the interval; if the battery power falls below 0.0%, we stop the simulation with a message:

```
while True:
    # update the location of the car
    self.LOCATION += distanceTraveled

    # how much battery power left
    batteryLeft = self.BATTERY_LEVEL \
        * self.BATTERY_CAPACITY - batteryUsed

    # update the level of the battery
    self.BATTERY_LEVEL = batteryLeft \
        / self.BATTERY_CAPACITY

    # if we run out of power -- stop
    if self.BATTERY_LEVEL <= 0.0:
        break
```

If we are still driving, we look up the nearest two recharge `stations (...)` and check the distance to the nearest one:

```
        # along the way -- check the distance to
        # the next two recharge stations
```

```
nearestRechargeStations = \
    [gs for gs in self.rechargeStations
        if gs.LOCATION > self.LOCATION] [0:2]

distanceToNearest = [rs.LOCATION \
    - self.LOCATION
    for rs in nearestRechargeStations]

# are we currently passing a recharging station?
passingRechargeStation = self.LOCATION \
    + distanceTraveled > \
        nearestRechargeStations[0].LOCATION

# will we get to the next one on the charge left?
willGetToNextOne = self.check(
    batteryLeft,
    nearestRechargeStations[-1].LOCATION)
```

If we are currently passingRechargeStation, we have the choice of either pulling over or seeing if we willGetToNextOne. If we are passing a recharge station and have determined that we will most likely not make it to the next one, we stop driving and start the recharging process:

```
if passingRechargeStation \
    and not willGetToNextOne:

    # the charging can be interrupted by the
    # driver
    try:
        # how long will it take to fully recharge?
        timeToFullRecharge = \
            (1 - self.BATTERY_LEVEL) \
            / nearestRechargeStations[0] \
                .RECHARGE_SPEED

        # start charging
        charging = self.env.process(
            self.charging(timeToFullRecharge,
                nearestRechargeStations[0] \
                .RECHARGE_SPEED))

        # and see if the driver will drive off
        # earlier than the car is fully recharged
        yield self.env.process(self.driver \
            .drive(self, timeToFullRecharge))
```

```
            # if the he/she does -- interrupt charging
            except simpy.Interrupt:
                charging.interrupt()
        # and wait for the next update
        yield self.env.timeout(interval)
```

Notice how we are wrapping our code within the try-except clauses; within the try statement, we first determine `timeToFullRecharge` and begin the recharge process. However, at the same time, we call the `Driver` object (specifically, the `drive(...)` method) and start the process that will determine the fate of our recharge efforts—are we going to be interrupted or not?

Assuming that we are not interrupted, the code within the try clause in the `charging(...)` method will execute fully. Note that we charge the battery incrementally every second of the simulation just in case we are interrupted somewhere in between:

```
    def charging(self, timeToFullRecharge, rechargeSpeed):
        # we are starting the recharge process
        try:
            for _ in range(int(timeToFullRecharge)):
                self.BATTERY_LEVEL += rechargeSpeed
                yield self.env.timeout(1)

        except simpy.Interrupt:
            pass
```

If the process gets interrupted by the Driver, the exception caught within the `driving(...)` method will be cascaded to `charging(...)`. The exception is then passed without doing anything but stopping the charging process and returning the execution to `driving(...)`.

Once you execute the script, you will see something as follows:

```
----------------------------
Time     Batt.    Dist.
----------------------------
0        0.87     12.2
15       0.80     24.4
30       0.74     36.5
45       0.67     48.7
60       0.61     60.9
75       0.54     73.0
90       0.48     85.2
----------------------------
Charging at 105
Fully charged...
----------------------------
```

If the charging is interrupted, a message will be printed:

```
1046      0.61       620.9
1061      0.54       633.1
1076      0.48       645.3
1091      0.41       657.4
1106      0.35       669.6

----------------------------

Charging at 1121
Charging interrupted at 1189

----------------------------

1189      0.97       681.8
1204      0.90       694.0
1219      0.84       706.1
```

If, during one of the runs of the simulation, the car runs out of power, you will see something similar to the following:

```
32764     0.05      17689.4
32779     0.01      17699.8


!~!~!~!~!~!~!~!~!~!~!~!~!~!~!~
RUN OUT OF JUICE...
!~!~!~!~!~!~!~!~!~!~!~!~!~!~!~
```

Determining if a population of sheep is in danger of extinction due to a wolf pack

One of the most famous agent-based simulations is the sheep-wolf predation example.

The model simulates two populations of animals: sheep and wolves coexisting together on a plane. Sheep move around and eat grass that grows on the plane; eating grass gives the sheep energy. Wolves predate sheep and that is how they get their energy. To move around the plane costs the animal some energy. If any animal's energy falls below 0, it dies.

In this recipe, we will build a 300-by-300 plane (grid) and populate it with 6,000 sheep and 200 wolves (initially). We will also introduce the concept of inheritance: we will create a generic `Animal` class and then derive `Sheep` and `Wolf` classes from it. The idea behind it is simple: as the animals share some common characteristic (that is, both of them need to move around the plane), we do not have to implement the same code in two places of the script.

Getting ready

To execute this recipe, you will need NumPy and SimPy. No other prerequisites are required.

How to do it...

The code for this recipe is lengthy but should be fairly straightforward to follow (the sim_ sheepWolvesPredation.py file):

```python
import numpy as np
import simpy
import collections as col

if __name__ == '__main__':
    # create the environment
    env = simpy.Environment()

    # create the plane
    plane = Plane(env, LAND, GRASS_COVERAGE,
        INITIAL_SHEEP, INITIAL_WOLF)

    # and run the simulation
    env.run(until = SIM_TIME)
```

How it works...

As before, we start with creating .Environment().

Next, we create the Plane object that our animals will be grazing on (the following code has been abbreviated for brevity):

```python
class Plane(object):
    def __init__(self, env, bounds, grassCoverage,
        sheep, wolves):

        # generate the grass and animals
        self.generateGrass()
        self.generateSheep()
        self.generateWolves()

        # and start monitoring and simulation processes
        self.monitor = self.env.process(
            self.monitorPopulation())
        self.action = self.env.process(self.run())
```

The `Plane` class generates the plane and plants the grass first; we use the `.generateGrass()` method to achieve this:

```
def generateGrass(self):
    # number of tiles on the plane
    totalSize = self.bounds[0] * self.bounds[1]

    # how many of them will have grass
    totalGrass = int(totalSize * self.grassCoverage)

    # randomly spread the grass on the plane
    grassIndices = sorted(
        choice(totalSize, totalGrass, replace=False))

    for index in grassIndices:
        row = int(index / self.bounds[0])
        col = index - (self.bounds[1] * row)

        self.grass[row][col] = 1
```

The method, based on the simulation parameters, checks how many tiles in total are on the plane and determines how many of them should be populated with grass. We then use the `.random.choice(...)` method of `NumPy` to randomly select the indices of the plane that will be growing grass. Finally, we place the grass in the right spots on the `Plane`.

Once the grass is generated, we then generate animals:

```
def generateSheep(self):
    # place the sheep randomly on the plane
    for _ in range(self.noOfSheep):
        pos_x = rint(0, LAND[0])
        pos_y = rint(0, LAND[1])
        energy = rint(*ENERGY_AT_BIRTH)

        self.sheep.append(
            Sheep(
                self.counts['sheep']['count'],
                self.env, energy, [pos_x, pos_y], self)
            )
        self.counts['sheep']['count'] += 1
```

Starting with sheep, we start by deciding the primary attributes of the `Sheep` on the plane: where it will be positioned and how much initial energy it will have. Once this is decided, we simply append the new animal to the `self.sheep` list.

You can create a loop in Python without creating the iterating index such as i if you do not need it; simply use _ instead.

The population of wolves is generated in a very similar manner:

```python
def generateWolves(self):
    # place the wolves randomly on the plane
    for _ in range(self.noOfWolves):
        pos_x = rint(0, LAND[0])
        pos_y = rint(0, LAND[1])
        energy = rint(*ENERGY_AT_BIRTH)

        self.wolves.append(
            Wolf(
                self.counts['wolves']['count'],
                self.env, energy, [pos_x, pos_y], self)
            )

        self.counts['wolves']['count'] += 1
```

Both of the `Sheep` and `Wolf` classes are derived from the `Animal` class; both are animals after all:

```python
class Animal(object):
    def __init__(self, i, env, energy, pos, plane):
        # attributes
        self.energy = energy
        self.pos = pos           # current position

        # is the animal still alive
        self.alive = True
        self.causeOfDeath = None

        # when did the animal ate the last
        self.lastTimeEaten = 0

        # range of movements
        self.movements = [i for i in range(-50,51)]

        # pointer to environment and the plane
        self.env = env
        self.plane = plane
        self.id = i
```

The `Animal` class controls all the main (and common across `Sheep` and `Wolf`) characteristics: lets the animal `move(...)` (and returns its position) and allows the animal to `die(...)` and check the cause of death—you can check whether the animal `isAlive(...)` and see how much energy the animal has left. As you can see, the preceding characteristics are common for all animals. The `Animal` class also stores all the common attributes of an animal: its position, amount of energy amassed, or whether the animal is still alive among others.

To `move(...)` the animal on the plane, we randomly select the movement along the horizontal and vertical axes and move the animal in that direction:

```
def move(self):
    # determining the horizontal and vertical moves
    h = choice(self.movements)
    v = choice(self.movements)

    # adjusting the position
    self.pos[0] += h
    self.pos[1] += v

    # making sure we do not go outside the predefined land
    self.pos[0] = np.min(
        [np.max([0, self.pos[0] - 1]), LAND[0] - 1]
    )

    self.pos[1] = np.min(
        [np.max([0, self.pos[1] - 1]), LAND[1] - 1]
    )

    # and subtracting the energy due to move
    self.energy -= (h+v) / 4
```

The method makes sure that you do not cross the bounds of the plane on your moves. It also removes a portion of the energy due to moving.

What differs between `Sheep` and `Wolf` is what they eat—sheep `eatGrass(...)` while wolves `eatSheep(...)`:

```
def eatGrass(self):
    '''
        Sheep eat grass
    '''
    if self.plane.hasGrass(self.pos):
        # get the energy from grass
        self.energy += ENERGY_FROM_GRASS
        self.lastTimeEaten = self.env.now
```

```
# and flag that the grass has been eaten
self.plane.grassEaten(self.pos)

if self.energy > 200:
    self.energy = 200
```

To eat grass, we first check whether the current position of the sheep on the plane is populated with grass. If it is, we increase the energy of the sheep by the amount specified in the ENERGY_FROM_GRASS variable. We also specify that sheep's energy level cannot exceed 200; after all, we cannot grow indefinitely. We also update the time since the sheep has eaten last. We will use this to determine the chances of an animal to reproduce; the idea is that if you haven't eaten for a long time, you are less likely to reproduce.

> Note that when we derived Sheep (or Wolf) from the Animal class, we still referred to self (for example, self.energy) even though all the logic is implemented within the parent (Animal) class.

In contrast, the eatSheep(...) method first gets all the Sheep at the current position of the wolf and decides how many of the sheep the wolf will eat:

```
def eatSheep(self):
    # get the sheep at the particular position on the
    # plane
    sheep = self.plane.getSheep(self.pos)

    # decide how many will be eaten
    howMany = np.random.randint(1,
        np.max([len(sheep), 2]))

    # and feast
    for i, s in enumerate(sheep):
        if s.isAlive() and i < howMany:
            self.energy += s.getEnergy() / 20
            s.die('eaten')

    if self.energy > 200:
        self.energy = 200

    # update the time of the last meal
    self.lastTimeEaten = self.env.now
```

We then loop through all the Sheep objects, check whether the sheep is still alive, and, if it is, the wolf eats it. The wolf gets a portion of the energy that the sheep had and is also bound at 200 units of energy.

Going back to the `Plane` object, once all the animals have been generated, we start the simulation processes.

The first process to initiate is `monitorPopulation(...)`:

```
def monitorPopulation(self):
    # the process checks for animals that run out of
    # energy and removes them from simulation
    while True:
        for s in self.sheep:
            if s.energy < 0:
                s.die('energy')

        for w in self.wolves:
            if w.energy < 0:
                w.die('energy')

        # clean up method
        self.removeAnimalsThatDied()

        yield self.env.timeout(1)
```

The method loops through all the animals (both `Sheep` and `Wolf`) and checks for the animal's energy level; if it is below `0`, the animal dies. Once all the animals have been flagged to die, the `removeAnimalsThatDied(...)` method is called:

```
def removeAnimalsThatDied(self):
    '''
        Clean up method for removing dead animals
    '''
    # get all animals that are still alive and those
    # that died
    sheepDied = []
    wolvesDied = []

    sheepAlive = []
    wolvesAlive = []

    for s in self.sheep:
        if s.isAlive():
            sheepAlive.append(s)
        else:
            sheepDied.append(s)

    for w in self.wolves:
```

```
        if w.isAlive():
            wolvesAlive.append(w)
        else:
            wolvesDied.append(w)

    # keep only those that are still alive
    self.sheep = sheepAlive
    self.wolves = wolvesAlive

    # and finally -- release the memory by deleting the
    # animal objects
    for s in sheepDied:
        del s

    for w in wolvesDied:
        del w
```

The method loops through all the animals and separates the animals that are still alive from those that died; those that survived continue to live through another iteration. Finally, we delete the animal objects so that they do not take up memory.

Another process that starts up is the main simulation loop, the run(...) method. The method controls the flow of the simulation:

```
def run(self):
    '''
        Main loop of the simulation
    '''
    while True:
        # first, move the animals on the plane
        self.updatePositions()

        # and let them eat
        self.eat()

        # then let's see how many of them will reproduce
        self.reproduceAnimals()

        # and keep track of the grass regrowth
        self.env.process(self.regrowGrass())

        # and wait for another iteration
        yield self.env.timeout(1)
```

During each iteration of the simulation, all the animals are allowed to move around the plane (a single move). The `updatePositions(...)` method loops through all the animals and calls the `move(...)` method:

```
def updatePositions(self):
    for s in self.sheep:
        s.move()

    for w in self.wolves:
        w.move()
```

Next, each animal is allowed to `eat(...)`. Similarly, the method loops through all the `Sheep` objects and instructs the animal to `eatGrass(...)`; all the wolves are instructed to `eatSheep(...)`, if there are any:

```
def eat(self):
    for s in self.sheep:
        s.eatGrass()

    for w in self.wolves:
        w.eatSheep()
```

Once the animals have eaten, the simulation will `reproduceAnimals(...)`:

```
def reproduceAnimals(self):
    # reproduce sheep
    for s in self.sheep:
        # determine if the animal will reproduce
        willReproduce = np.random.rand() < \
            (SHEEP_REPRODUCE * 3 / \
                (self.env.now - s.lastTimeEaten + 1))

        # if will reproduce and is still alive --
        # give birth at the same position as the mother
        if willReproduce and s.isAlive():
            energy = rint(*ENERGY_AT_BIRTH)
            self.sheep.append(
                Sheep(self.counts['sheep']['count'],
                    self.env, energy,
                    s.getPosition(), self))

            # increase the overall count of sheep
            self.counts['sheep']['count'] += 1

    # reproduce wolves
    for w in self.wolves:
```

```
        # determine if the animal will reproduce
        willReproduce = np.random.rand() < \
            ( WOLF_REPRODUCE / \
                (self.env.now - w.lastTimeEaten  + 1))
        # if will reproduce and is still alive --
        # give birth at the same position as the mother
        if willReproduce and w.isAlive():
            energy = rint(*ENERGY_AT_BIRTH)
            self.wolves.append(
                Wolf(self.counts['wolves']['count'],
                    self.env, energy,
                    w.getPosition(), self))

            # increase the overall count of wolves
            self.counts['wolves']['count'] += 1
```

The method loops through all the animals and decides at random if the animal will reproduce or not. The probability of producing an offspring for an animal is dependent (as mentioned earlier) on the time that the animal has eaten last: the longer it has been for the animal to eat, the lower the chances that it will reproduce. If it is decided that the animal willReproduce and isAlive, a new Sheep or Wolf is created in the same position as its mother.

Having increased the population, we finally start the process of regrowing the grass:

```
    def regrowGrass(self):
        # time to regrow the grass
        regrowTime = 2
        yield self.env.timeout(regrowTime)

        # then we make the grass available at the position
        for pos in self.grassEatenIndices[
            self.env.now - regrowTime]:
            self.grass[pos[0]][pos[1]] = 1
```

The regrowGrass(...) method loops through all the positions that have been recently marked as grazed on and regrows the grass after a specified amount of time. Each time a Sheep object eats grass, the index in the .grass variable is set to 0 in the position that the sheep grazed, and the position is also added to the .grassEatenIndices defaultdict object that keeps track of when each position was grazed on. This allows you to alter regrowTime and see how this affects the population of sheep.

This finalizes a single iteration of the simulation. When you execute the script, you should see something as follows:

		Sheep		Died		Wolves		Died
T	Live	Born	Energy	Eaten	Live	Born	Energy
0	6858	858	0	0	201	1	0
1	6971	529	388	28	179	1	23
2	6894	454	498	33	161	1	19
3	6814	417	458	39	149	2	14

The T denotes the time of the simulation, Live specifies the number of animals alive, Born shows the number of births, and Died gives a split of how many animals died—either due to lack of energy or (in the case of Sheep) being eaten.

Index

Symbols

.fillna(...) method
 URL 37
.get_dummies(...) method
 URL 42
.KNeighborsClassifier(...) method
 URL 154
.supervised.trainers module
 URL 94

A

agent-based simulations
 limitations 330
 reference link 330
 sheep-wolf predation example 348-358
alternative specific constants (ASCs) 313
American River dataset
 URL 198
Anaconda
 URL 3
**Application Programming Interface
 (API) 141, 268**
arpack
 reference link 137
article
 topic, handling 289-292
Artificial Neural Networks (ANN)
 about 92
 reference link 99
autocorrelation function (ACF)
 about 203
 reference link 204
auto-regressive (AR) component 204

**autoregressive integrated moving average
 (ARIMA) model**
 about 223
 future, forecasting with 223-232
autoregressive moving average (ARMA) model
 about 223
 future, forecasting with 223-232
Axes object
 URL 55

B

Baxter-King (BK) filter
 about 215
 URL 215
BIRCH algorithm
 reference link 125
 used, for searching groups of potential
  subscribers 123-125
Bokeh
 URL 61
bottom up approach 119

C

calls
 classifying, with decision trees 83-88
 classifying, with neural networks 92-99
CART
 reference link 181
 used, for estimating output of
  electric plant 177-181
categorical variables
 encoding 41, 42

Natural Language Toolkit (NLTK)
about 274
URL 297
NetworkX
about 234
URL 244
neural networks
reference link 196
used, for classifying calls 92-99
used, for predicting output of
    power plant 194-196
n-grams
handling 282
non-square matrices
reference link 162
noun phrases (NP) 292
NumPy
URL 40

O

observations
binning 39, 40
obsIter 314
online converters
URL 294
OpenRefine
data, exploring with 28-30
data, opening from 25-28
data, transforming from 25-28
URL 26
Optunity
URL 191
Ordinary Least Squares (OLS)
about 172
reference link 177
used, for forecasting 172-177
out-of-energy occurrences
simulating, for electric car 342-348

P

pandas
HTML pages, retrieving with 16-18
URL 3
parsers
URL 271

partial autocorrelation function (PACF)
about 203
reference link 204
parts of a sentence
URL 295
parts of speech
identifying 282-289
tags, URL 285
piping
about 79
URL 79
PostgreSQL database
URL 20
potential subscriber groups
searching, with BIRCH algorithm 123-125
searching, with DBSCAN algorithm 123-125
prepositional phrases (PP) 292
Principal Components Analysis (PCA)
about 131
comparing, with ICA and truncated
    SVD 151-156
reference link 133
URL, for visualization 135
using 135-140
principal components (PCs)
about 128, **164**
reference link 180
searching, randomized PCA used 140-146
three-dimensional scatter plots,
    creating 128-130
probability density function (PDF) 55
Pseudo-F 103
psycopg2
URL 20
PyBrain
installing 93
URL, for modules 94
PyMongo
URL 24
Python
CSV/TSV files, reading with 2-7
CSV/TSV files, writing with 2-7
date objects, handling 198-203
excel files, reading with 9-11
excel files, writing with 9-11
JSON files, reading with 8, 9
JSON files, writing with 8, 9

www.ingramcontent.com/pod-product-compliance
Lightning Source LLC
Chambersburg PA
CBHW062046050326
40690CB00016B/3001